Manual of Water Supply Practices

5

# Geomembrane Floating Cover and Liners

## or Potable-Water Reservoirs

Fourth Edition

merican Water Works
ssociation

Manual of Water Supply Practices

M25

# Geomembrane Floating Covers and Liners for Potable Water Reservoirs

Fourth Edition

American Water Works Association

Manual of Water Supply Practices—M25, Fourth Edition

# Geomembrane Floating Covers and Liners for Potable Water Reservoirs



Senior Editorial Manager – Book Products: Melissa Valentine
Senior Manuals Specialist: Molly Beach
Manager – Publishing Operations: Gillian Wink
Copyright and Permissions Manager: Peggy Tyler
Technical Editor: Dianne Beirne
Cover Composition: Michael Labruyere
Production: Innodata

**Library of Congress Cataloging-in-Publication Data**

Names: American Water Works Association.
Title: M25 - geomembrane floating covers and liners for potable water reservoirs / by American Water Works Association.
Other titles: Flexible-membrane covers and linings for potable water reservoirs | Geomembrane floating covers and liners for potable water reservoirs
Description: Fourth edition. | Denver, CO : American Water Works Association, [2023] | Revised edition of: Flexible-membrane covers and linings for potable water reservoirs, 3rd ed. Denver, CO : American Water Works Association, [2000]. | Includes bibliographical references and index. | Summary: "This manual is a technical reference guide for designing, installing, operating, and maintaining geomembrane liners and floating covers for potable water storage reservoirs"-- Provided by publisher.
Identifiers: LCCN 2023032037 (print) | LCCN 2023032038 (ebook) | ISBN 9781647171438 (paperback) | ISBN 9781613006740 (adobe pdf)
Subjects: LCSH: Reservoirs--Linings--Handbooks, manuals, etc. | Membranes (Technology)--Handbooks, manuals, etc.
Classification: LCC TD395 .F59 2023 (print) | LCC TD395 (ebook) | DDC 628.1--dc23/eng/20230821
LC record available at https://lccn.loc.gov/2023032037
LC ebook record available at https://lccn.loc.gov/2023032038

Printed in the United States of America

ISBN 978-1-64717-143-8 eISBN 978-1-61300-674-0



American Water Works Association

American Water Works Association
6666 West Quincy Avenue
Denver, CO 80235-3098
303.794.7711
awwa.org

# Contents

This page intentionally blank.

# Figures

This page intentionally blank.

# Acknowledgments

This manual was prepared by the AWWA Standards Committee on Geomembrane Floating Covers and Geomembrane Liners for Potable Water Storage. The membership of the committee at the time it approved this manual was:

Douglas A. Hilts, *Chair*

*General Interest Members*

D.H. Gerber, Gerber Consulting, Carmel, Calif.
P.F. Greiner, NSF International, Ann Arbor, Mich.
D.A. Hilts, Hilts Consulting Group Inc., Yorba Linda, Calif.
D. Mason (*liaison, nonvoting*), Standards Council Liaison, O&M Enterprises, Holden, Mo.
C.W. Neal II, Broken Arrow, Okla.
P.J. Olson (*liaison, nonvoting*), Standards Engineer Liaison, AWWA, Denver, Colo.
I.D. Peggs, I-Corp International Inc., Lake Worth, Fla.
W.A. Way, Consulting Engineer, Johnstown, Colo.

*Producer Members*

S. Falk, Conterra, Inc., El Cajon, Calif.
D.J. McCullough, Colorado Lining International Inc., Mission Viejo, Calif.
W.D. Shehane, Seaman Corporation, Wooster, Ohio
C. Taylor, Carlisle SynTec Inc., White Salmon, Wash.

*User Members*

T.P. Campbell, Metropolitan Water District of Southern California, Los Angeles, Calif.
D.G. Miller, Manchester Water Works, Manchester, N.H.
J.T. Otoshi, Los Angeles Water & Power, Los Angeles, Calif.

This page intentionally blank.

Chapter 1

# Introduction

## PURPOSE AND SCOPE

This manual is intended to serve as a technical reference guide for designing, installing, operating, and maintaining geomembrane liners and floating covers for potable water storage reservoirs. Source information was gathered by reviewing and collecting the experience of designers, manufacturers, and owners of geomembrane products available to the water industry.

This manual is not intended to be a design handbook or the equivalent of an AWWA standard. Rather, it should serve as a technical reference document to be used in conjunction with ANSI/AWWA D130, Geomembrane Materials for Potable Water Applications. The purpose of this manual is to recommend minimum design criteria and minimum operations and maintenance procedures for potable water reservoir floating covers and liners. Site-specific conditions should be evaluated by a qualified professional engineer and the recommendations adjusted accordingly to provide a comparable performance level.

Information in this manual may not apply to all types of installations or materials. In applying specific recommendations, the user must assume responsibility for accommodating a specific set of conditions.

## DESCRIPTION OF FACILITIES

Geomembrane liners and floating covers can be incorporated into many types of water-storage facilities, both new construction and rehabilitated structures. Some examples of facilities include:

- **An earthen reservoir for storage of untreated or raw water.** A geomembrane liner used in this type of installation is normally intended to minimize water losses due to seepage and prevent the introduction of any contaminants that may exist in the underlying soil. In arid regions, a geomembrane floating cover might be installed to reduce water loss due to evaporation as well.

- **An earthen reservoir for storage of potable water.** Such a facility might be located at a water treatment plant or within a distribution system. Normally, the structure would incorporate both a liner and a floating cover to reduce water losses and to protect the treated water from contamination.
- **A storage facility that has developed an excessive leakage condition.** This type of structure might be constructed of steel, concrete, or another material, or it might be asphalt-lined. Normally, a geomembrane liner would be installed without a floating cover to reduce leakage and protect the stored water from outside contamination.
- **An uncovered storage facility requiring protection from contamination.** A geomembrane floating cover can quite often be used to meet these requirements. Many large, open reservoirs have been covered in this manner.

## ADVANTAGES AND DISADVANTAGES

The principal advantages of geomembrane liners and floating covers are:

- prevention of airborne and groundwater contamination,
- prevention of evaporation and seepage losses,
- reduction of algal growth,
- reduction of disinfection chemical demand,
- low capital cost, and
- superior seismic performance compared to more traditional structural applications.

The principal disadvantages of geomembrane liners and floating covers are:

- susceptibility to damage by ice action and vandalism,
- floating covers having shorter life expectancy than rigid covers, and
- the need for regularly scheduled maintenance to inspect the geomembrane material and system integrity, remove surface water and debris, and make repairs.

## PERMEATION

Selection of materials is critical for water-storage facilities and distribution piping in locations where the reservoir materials and/or pipe will likely be exposed to significant concentrations of pollutants composed of low-molecular-weight petroleum products or organic solvents and their vapors. Research has documented a potential for such compounds to permeate through plastic pipe materials such as polyethylene, polybutylene, polyvinyl chloride (PVC), and asbestos cement and through elastomers such as those used in jointing gaskets and packing glands. If a reservoir must be located in, or a water pipe must pass through, an area subject to such contamination, consult with the manufacturer regarding permeation of pipe walls, jointing materials, reservoir geomembranes, and similar appurtenances before selecting materials for use in that area.

## DEFINITIONS

For the purpose of this manual, definitions are provided as follows. For additional definitions, the reader may reference ASTM D4439, Standard Terminology for Geosynthetics.

**Access features:** Access hatches/pathways and/or perimeter access vaults incorporated into a floating cover to allow access by personnel under the floating cover while it is floating or inflated.

**Air vents:** Air vents are provided to release entrained air that enters the reservoir with the water and subsequently accumulates beneath the floating cover. Air vents are also used to prevent formation of vacuum in areas where water exits the reservoir (e.g., outlets, overflow drains). Air vents can have a negative effect on wind resistance of floating covers, so their use should be minimized. Air vents are used on geomembrane liners to prevent uplift due to gas accumulation or air pushed ahead of rising groundwater levels.

**Appurtenances:** Ancillary items that are installed on or under the geomembrane liner or floating cover to permit their proper operation or function.

**Bridging:** Unsupported geomembrane material spanning between areas of support.

**Closed-cell float:** (*See* Floats)

**Coating:** A film built up through successive applications of an appropriate dispersion or solution system to achieve a specific dry-film thickness. Generally, coatings are applied to steel tanks for the purpose of protecting the steel against corrosion.

**Contractor:** The party under contract with the owner to furnish a completed geomembrane floating cover or liner.

**Cut-and-fill reservoir:** A water-holding facility constructed by excavating a central portion and, normally, forming the excavated earth as fill material into an embankment that increases the capacity of the reservoir while avoiding costs of disposal for the excavated material.

**Earthen reservoir:** (*See* Cut-and-fill reservoir)

**Engineer:** The party responsible for the technical design and preparation of design drawings and technical specifications for the geomembrane liner and/or floating cover.

**Fabricator:** The party that joins rollstock strips of geomembrane material into large panels in a fabrication plant for delivery to the jobsite for installation. The fabricator may also shop fabricate liner and/or floating cover accessory components.

**Flexible baffles:** Vertical curtains of geomembrane material, usually the same material as the geomembrane floating cover, typically extending from the water surface to the reservoir floor. Flexible baffles may be attached to the reservoir floating cover; rigid cover when present; or structurally supported by some other independent system. Flexible baffles direct the flow of water throughout the area of the reservoir to lengthen the contact/residence time of disinfected water.

**Floating cell:** An area of floating cover bounded by surface water collection troughs.

**Floating cover:** A continuous, flexible, tensioned, impervious, and weather-resistant geomembrane material that floats on the surface of a water reservoir to function as a floating cover system (see Figure 1-1). Air-supported floating cover systems, bladder tanks, and nontensioned floating covers are not discussed in this manual. (Nontensioned floating covers are no longer considered state-of-the-art technology.)

**Floats:** Closed-cell, semi-rigid foam members, usually encased in the same type of geomembrane material as the floating cover and permanently joined to it. Floats are usually required for stabilization, to facilitate surface water removal, to frame access and other hatches, and to provide buoyancy.

**Geomembrane:** Any of several continuous, flexible, and impermeable geomembranes, usually ranging in thickness between 0.036 in. (0.91 mm) and 0.060 in. (1.52 mm), which may be fabricated into large panels for final assembly in the field.

The geomembrane may be nonreinforced or scrim-reinforced, depending on the physical requirements of the application. Three types of geomembrane materials are generally used: plastic (thermoplastic); cured elastomer; and uncured elastomer (thermoplastic elastomer).

**Geomembrane floating cover:** (*See* Floating cover)

**Geomembrane manufacturer:** The party that produces a flexible geomembrane rollstock material, which is used to create geomembrane panels and other liner and/or floating cover accessories.

**Geotextile:** A permeable synthetic fabric, typically nonwoven, used in reservoirs to protect a geomembrane liner from damage due to rough or uneven surfaces.

**Grillage:** A framework installed under the floating cover to protect the floating cover under certain reservoir operating conditions. Grillage is commonly a three-dimensional framework installed over structures or openings in the reservoir floor or slopes that the floating cover must be supported on in dewatered or low operating conditions.

**High operating level (HOL):** The maximum water surface elevation at which the reservoir is commonly operated.

**Inflation features:** Specific design features that allow a floating cover to be conveniently, safely, and temporarily inflated and accessed for reservoir inspection, cleaning, and maintenance. Inflation should be performed when the reservoir is empty.

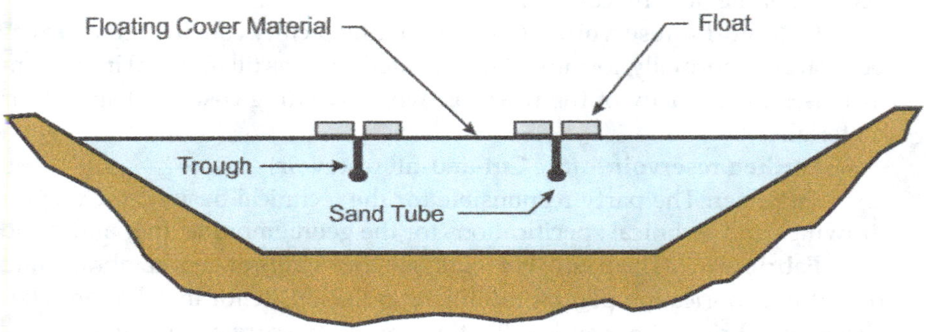


**(a) Schematic of weight-tensioned floating cover with full reservoir (not to scale)**

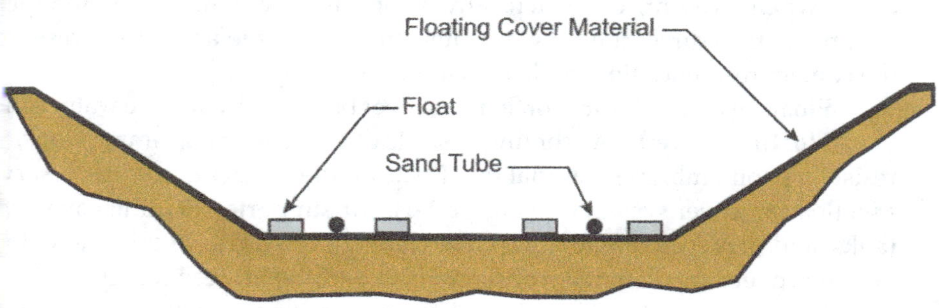


*Source:* Metropolitan Water District of Southern California

**(b) Schematic of weight-tensioned floating cover with empty reservoir (not to scale)**

**Figure 1-1 Cross sections of a typical weighted-sump floating-cover system, shown full and empty**

**Inspector:** The party responsible for performing inspection tasks associated with the manufacturing, shop fabrication, and/or field installation of a geomembrane liner and/or floating cover.

**Installer (installation contractor):** The party that installs the geomembrane liner and/or floating cover in the field.

**Liner:** A reservoir component designed to prevent leakage of water or infiltration of contaminants into the contained water. This component may be of geomembrane material, steel, reinforced concrete, asphalt, compacted earth, or a combination of these materials. Some designs integrate geomembrane liner functions with the structural members of the reservoir, such as concrete walls and slabs, or asphalt liners.

**Low operating level (LOL):** The minimum water surface elevation at which the reservoir is commonly operated.

**Normal operating level (NOL):** The water surface elevation or range at which the reservoir is operated under normal conditions.

**Owner:** The party contracting for the installation of the floating cover or liner and also responsible for the ongoing operation and maintenance of the reservoir, liner, and floating cover.

**Panel:** The product of the assembly of two or more rollstock strips of flexible-geomembrane material cut and assembled by the fabricator using factory-seaming techniques to form a portion of a geomembrane liner or floating cover ready for field assembly by the installer.

**Project documents:** Project specifications and project drawings prepared by a licensed professional engineer and tailored to describe unique project-specific requirements. Project specifications typically include text and technical drawings.

**Purchaser:** The party that purchases any materials or work to be performed.

**Reinforcement:** (*See* Scrim)

**Repair float:** A float, unattached to the cover, used to temporarily elevate a portion of floating cover geomembrane material above the water surface to make repairs.

**Reservoir:** A pond, lake, basin, or other water impoundment facility, either naturally formed or man-made, which is used for storage, regulation, and control of water. (*See also* Tank)

**Rollstock:** Rolls of flexible geomembrane material, as produced by the manufacturer, usually in widths of 5–15 ft.

**Scrim:** Reinforcing material or fabric encapsulated within the flexible geomembrane rollstock material.

**Scrim edge:** The edge of the reinforcing material encased in a geomembrane.

**Substrate:** The soil, concrete, asphalt, or other surface that is below and supports the geomembrane liner.

**Sump:** (*See* Surface water collection trough)

**Supplier:** (*See* Fabricator; Geomembrane manufacturer)

**Surface water collection trough:** A portion of a weight-tensioned floating cover where reservoir geometry determines placement of excess material that distributes itself automatically and uniformly into a fold or sump configuration that collects and conveys surface water to the surface water removal system.

**Surface water removal system:** A series of components that is responsible for the removal of surface water from on top of the floating cover. The surface water removal system typically consists of pumps or gravity drains.

**Tank:** A reservoir created in whole by constructing an engineered structure in which water or other liquids are stored. Ponds and lakes, naturally or artificially constructed, are not classified as tanks (*see also* Reservoir). Several general construction materials are: steel, reinforced concrete, prestressed concrete, wood, fiberglass-reinforced plastic, and combinations of these materials.

*Source:* Courtesy of Layfield USA

**Figure 1-2 Mechanically tensioned floating cover**

**Tensioning features:** Floating covers are often fitted with flexible weights or other devices for maintaining floating cover tension. These features impart a discrete, low-level tension to the floating cover geomembrane that stabilizes the floating cover system and provides for improved surface water runoff and personnel access (Figure 1-2).

## SURVEY OF EXISTING FACILITIES

Surveys of existing storage facilities were conducted in 1979 and 1994 to develop technical information and performance records of various types of geomembrane materials and designs. A questionnaire was developed and sent to known owners of tanks and reservoirs using geomembrane liners or floating covers. Results of the survey were incorporated into the content of this manual. Additional background information is available in published periodicals and technical journals. A limited bibliography is provided at the end of this manual.

Chapter 2

# Design and Installation of Geomembrane Floating Covers

## FLOATING COVER DESIGN

The water industry began using floating covers made of butyl rubber or ethylene propylene diene monomer (EPDM) synthetic rubber (and less commonly polyvinyl chloride [PVC]) during the mid-1960s. Since then, improvements in design, materials, and operating procedures have resulted in hundreds of successful floating cover installations throughout the world. Most of these installations are in the United States and Canada, and most have been constructed using reinforced geomembranes. Each geomembrane material has unique physical characteristics and requires special methods of construction and repair. These characteristics should be thoroughly evaluated before the geomembrane product is specified. Figures 2-1 through 2-7 illustrate several installations.

There are two basic types of floating covers: weight-tensioned and mechanically tensioned.

- Weight-tensioned floating covers are created with a series of weights and floats strategically located on the top of the floating cover. The weights serve to keep the floating cover material taut and take up excess material as the floating cover rises with the water surface. The fold or trough created by the weights serves as the surface water collection trough, which conveys surface water to the surface water removal system. The surface water removal system is located within the surface water collection troughs.

*Source:* Metropolitan Water District of Southern California

**Figure 2-1 Floating cover and geomembrane liner installed in 1998 on 525-mil-gal potable water reservoir in Monterey Park, Calif.**

*Source:* Metropolitan Water District of Southern California

**Figure 2-2 Floating cover and geomembrane liner installed in 2017 on 360-mil-gal potable water reservoir in Rolling Hills Estates, Calif.**

*Source:* Hilts Consulting Group

**Figure 2-3 Floating cover and geomembrane liner installed in 2001 on 275-mil-gal potable water reservoir in Mission Viejo, Calif.**

*Source:* Los Angeles Department of Water and Power

**Figure 2-4 Floating cover installed in 2014 on 83-mil-gal potable water reservoir in Eagle Rock, Calif.**

- *Mechanically tensioned floating covers* are created with a series of tower weights or tensioning springs located at the top of the slope around the reservoir perimeter. The tower weights or springs impart tension into the floating cover material to keep the floating cover taut. The excess material created as the floating cover rises is taken up by forming a fold at the top of the slope. The surface water removal system is located in the center portion of the floating cover. Some floating cover designs may be patented, requiring appropriate licenses.

*Source:* Hilts Consulting Group

**Figure 2-5 Floating cover installed in 2004 on 30-mil-gal potable water reservoir in Reno, Nev.**

*Source:* Los Angeles Department of Water and Power

**Figure 2-6 Floating cover installed in 2019 on 139-mil-gal potable water reservoir (Upper Stone Canyon Reservoir) in Los Angeles, Calif.**

## General Areas of Concern

The areas of concern that an owner must recognize are described in the following paragraphs.

**Engineer qualifications.** The drawings and specifications for a floating cover system should be prepared by a licensed engineer having specialized knowledge and experience in the design and operation of geomembrane floating cover systems.

*Source:* Los Angeles Department of Water and Power

**Figure 2-7 Floating cover installed on 67-mil-gal potable water reservoir (Lower Franklin Canyon Reservoir) in Los Angeles, Calif.**

Design drawings should include:

- floating cover surface area,
- overflow elevation,
- high operating level,
- normal operating range,
- low operating level,
- reservoir capacity,
- inlet and outlet flow capacities,
- floating cover material (type, thickness, color, reinforcing, etc.),
- seaming procedures,
- float layout,
- surface water removal system,
- hatches,
- air vents,
- perimeter attachment,
- penetrations, and
- other miscellaneous support material(s).

Design calculations should, at a minimum, address floating cover tension, entire floating cover buoyancy, hatch buoyancy, air vent buoyancy, and surface water removal systems.

**Economic consideration (budget limitation).** Floating covers compare favorably with other reservoir-covering alternatives in economic analysis based on initial construction cost. However, a life-cycle total cost comparison should consider expenses for periodic replacement and regularly scheduled maintenance, inspection, and repair over a time period consistent with the design life assumptions.

**Operations and maintenance methods and equipment.** Once a floating cover system is installed, access for maintenance and inspection to the interior of the reservoir must involve special precautions to keep the floating cover supported off the floor while workers are inside. Otherwise, provisions must be made for maintenance and inspection entry through floating cover access hatches while the reservoir is full.

**Jurisdictional and regulatory agency requirements for types of floating cover materials.** The floating cover material, chemical fusion agents, adhesives, and similar items should be suitable for contact with potable water. Materials should not contain any ingredients that, when dissolved in water, would exceed applicable water quality standards or add objectionable tastes or odors. Floating cover system materials should be checked to verify that they are approved by appropriate regulatory agencies and meet the requirements of ANSI/AWWA D130.

**Degree of reliability desired.** If the reservoir is a critical component of the water system, extensive out-of-service time for repairs and maintenance may not be acceptable. In such a case, a floating cover material with a proven record of satisfactory performance should be selected. In addition, the reservoir could be designed as a dual basin, so that one half can be taken out of service independently while the other half remains in use. Alternatively, a bypass system can be provided to facilitate the reservoir being taken out of service.

## Environmental Considerations

**Site selection.** The floating-cover design and layout should take into account the prevailing wind conditions at the site. Stabilizing floats and mechanical-tensioning devices should be positioned to minimize the effects of wind on the floating cover. Other special design features may be necessary to provide wind resistance. They should be considered where appropriate.

Reservoir side slopes should ideally be no steeper than 2 units horizontal to 1 unit vertical. Reservoirs with steeper side slopes should have special provisions to increase worker safety during construction and required maintenance and inspection activities.

Internal reservoir surfaces should be smooth, flat, and uniform. To the maximum extent possible, all inlet facilities, outlet facilities, overflows, and other structures within the reservoir should be flush with or below the plane of the reservoir liner. Grillages, grating, and other standoff structures should be low profile. Reservoir side slopes should be straight and uniform. Internal corners greater than 180 degrees should be avoided wherever possible.

On-site emergency power provisions with an automatic switchover should be considered to provide power to all surface water removal pumps and other vital equipment.

Other site considerations include bypass piping for the ability to isolate the reservoir without disrupting water delivery; presence of groundwater; and clearances from sewage facilities. The engineer should consult applicable state waterworks standards for regulatory requirements.

**Security.** Reservoirs may incorporate perimeter berms or other screening mechanisms to prevent visual observation of the floating cover from the surrounding terrain. Other security issues to consider include screening vegetation; nearby trees; debris; and animals.

The security fence should be placed far enough away from the floating cover to discourage rock throwing or other types of vandalism. The security fence should also be designed to prevent intrusion by animals that might damage the floating cover. Increased security measures, such as an intrusion alarm, video surveillance, or patrol personnel, may be necessary in areas subject to frequent vandalism. A second security fence around the immediate reservoir perimeter should be provided. The bottom of the reservoir perimeter fence should extend to within 2 in. of paving or perimeter curb.

**Climatic conditions.** The floating cover design must provide for prompt removal of surface water to minimize the possibility of cross-connection as well as surface water buildup and subsequent ice formation on top of the floating cover. Excessive amounts of surface water can also reduce the usable storage capacity of a reservoir and increase stress on the geomembrane by submerging floats.

In extreme cold-weather climates, wide fluctuations in reservoir levels can result in ice damage to geomembrane-type floating covers unless special design precautions are taken. Some precautions might include:

- *implementation of winter operation high- and low-level restrictions;*
- installation of an air-bubbler system or other device to prevent ice formation around projecting gate structures or within sumps;
- installation of a geomembrane chafing strip to prevent excessive abrasion of the floating cover at the inside top edge of the reservoir side slope;
- installation of submersible mixers to promote circulation and minimize ice buildup; and
- installation of other comparable measures to minimize ice buildup.

**Wind.** In general, wind uplift may not be an issue for properly designed floating covers during operations where the geomembrane material maintains continuous direct contact with the water surface. The introduction of air underneath the floating cover shall be avoided. Air may be entrained within the water entering the facility, or wind may enter underneath the floating cover through openings or inlets. The design shall include provisions to isolate those areas with the potential to allow air underneath the floating cover while still providing adequate venting. Overflow pipes or structures shall be equipped with provisions to prevent wind from entering underneath the floating cover by traveling up the overflow. Should air underneath the floating cover be unavoidable, the design shall account for wind uplift. Wind design speed may be determined from the local building official, historical wind speed records, or other rational method.

## Design Considerations

**Floating cover.** All floating covers should be tensioned to allow personnel easy and uninhibited access on top of the floating cover and to promote the removal of surface water. A floating cover design should provide for adequate tension and flotation to ensure stability of the entire floating cover surface sufficient for the safety of personnel walking on the floating cover for inspection, cleaning, and repair. Tension and flotation also promote proper drainage and collection of surface water. Floating cover tension should be greater than 0.50 lbs per linear inch but not greater than 3 percent of the parent material tensile strength.

For weight-tensioned floating covers, the surface water collection trough depth should not be greater than 6 ft. Floats should be positioned to provide the necessary buoyancy but also promote the flow of surface water into the surface water collection troughs. Buoyancy should be provided to facilitate all the required functions of the floating cover, including:

- surface water trough definition;
- surface water runoff enhancement;
- access pathways;
- air vent support;

- access hatch support;
- surface water removal equipment support;
- localized cover slack accommodation;
- cover chafing protection (over grillages, structures, overflow structures, inlets, outlets, pipes, etc.); and
- other unique areas.

Dedicated pathways may be provided on the floating cover. Pathways should be considered from the top of slope anchorage to, as a minimum, the low operating level. Pathways may be installed on top of the floating cover and may be created from textured slip-resistant material or steps formed of segmented foam floats encased in floating cover geomembrane material.

**Perimeter anchorage.** The floating cover should be attached above the overflow elevation with adequate freeboard around the entire reservoir perimeter and at all penetrations. A concrete anchor curb or anchor trench backfilled with compacted soil or concrete/slurry should be provided. The point of anchorage (curb batten bars or trench crest) shall be a minimum of 24 in. above the reservoir water surface elevation at overflow condition. The engineer should verify with regulatory agencies if additional freeboard is required. The attachment is typically made to the perimeter concrete anchor curb with corrosion-resistant fasteners, batten bars, and vulcanized rubber cushion strips.

**Hatches.** Hatches should be viewed as an extension of the floating cover and detailed with the intent to protect water quality. Hatches may be provided for access under the floating cover. Access underneath the cover should be considered for:

- divers and remotely operated vehicles (ROVs) when the reservoir is in service;
- personnel when the reservoir is empty and the floating cover is inflated;
- obtaining water quality samples;
- inserting repair floats; or
- other operational requirements.

A minimum net buoyancy of 2,000 lbs should be provided at all access hatches. Float height should be a minimum of 8 in. The floating cover geomembrane material should be attached to the top of floats a minimum of 8 in. above the nominal water surface. The hatch opening should be a minimum of 16 in. above the nominal water surface. The hatch opening should be a minimum of 3 ft by 3 ft. Hatches should be located at all key access points considering typical operations, inflation, and diver inspections. The number of hatches provided should be a minimum of 1 hatch per acre or 1 hatch per floating cell, whichever is greater. The connection of the floating cover to the hatch frame should be overlapped and gasketed to prevent insect intrusion. Hatches should be locked for security. See example access hatch in Figure 2-8.

**Air vents.** Air vents are used to remove air entrained in the water and underneath the floating cover but should be used with discretion and only as required. A minimum net buoyancy of 350 lbs should be provided at all air vents. Float height should be a minimum of 8 in. The floating cover geomembrane material should be attached to the top of floats. Air vents should be designed to maintain vent openings a minimum of 16 in. above water under all operating conditions and levels (Figure 2-9). They should shed water from rain or snowmelt onto the top surface of the floating cover. Air vents should be designed to inhibit the flow or intrusion of air from ambient wind. Otherwise, the floating cover may be inflated by ambient wind, potentially resulting in damage to the floating cover. Air vent openings should be directed downward not less than 45 degrees. Air vents should be fitted with a corrosion-resistant screen that complies with local and state regulations or other provisions to inhibit the intrusion of small animals and insects.

*Source:* Hilts Consulting Group

**Figure 2-8 Floating cover access hatch**

*Source:* Hilts Consulting Group

**Figure 2-9 Floating cover air vent**

*Source:* Hilts Consulting Group

**Figure 2-10 Floating cover vacuum vent**

Air vents may also serve as a vacuum break and should be provided at locations (e.g., outlet structures) where a vacuum may occur (Figure 2-10).

**Projections.** Walls, structures, and pipes protruding above the substrate should be avoided where a floating cover is to be installed. The floating cover must be free to rise and fall with the reservoir level without restriction or interference. If projections cannot be avoided, the design must include protection for the floating cover material when the reservoir is emptied to prevent puncture, chafing, tearing, or other damage caused by the projections.

**Surface water removal systems.** All floating covers should be designed with a surface water removal system (see examples in Figures 2-11, 2-12, and 2-13). There are several types of such systems. The selection of the system should be compatible with the type of floating cover and the availability of the owner's operations and maintenance personnel. The overall surface water removal system should include a primary and a secondary system. The primary surface water removal system should be permanent pumps or gravity drains. The primary system should be automated such that its operation or startup does not have to be manually initiated. The secondary surface water removal system may be permanent pumps, temporary pumps, gravity drains, or siphons. The secondary system does not need to be automated, but it should be capable of being deployed within a reasonable period if the primary system is not operational.

The primary surface water removal system should be designed based on a rational engineering approach. The system should have the capacity for the following:

- 10-year storm, 24-h rainfall duration, capacity to remove surface water 24 h after the storm has passed (48 h total), and
- 25-year storm, 24-h rainfall duration, capacity to remove surface water 48 h after the storm has passed (72 h total).

(a) Reservoir empty (b) Reservoir near full

*Source:* Hilts Consulting Group

**Figure 2-11 Floating cover rainwater collection trough**

*Source:* Hilts Consulting Group

**Figure 2-12 Floating cover inverted tee rainwater removal system**

The floating cover design should provide specific features for effective and continuous surface water collection and removal suited to site-specific conditions and requirements. Surface water should be removed as it is collected. In times of impending freeze, continuous removal of surface water from the floating cover is essential.

For pumped systems, no less than two pumps shall be provided. The total number of pumps provided should be based on the surface water removal rate with one pump

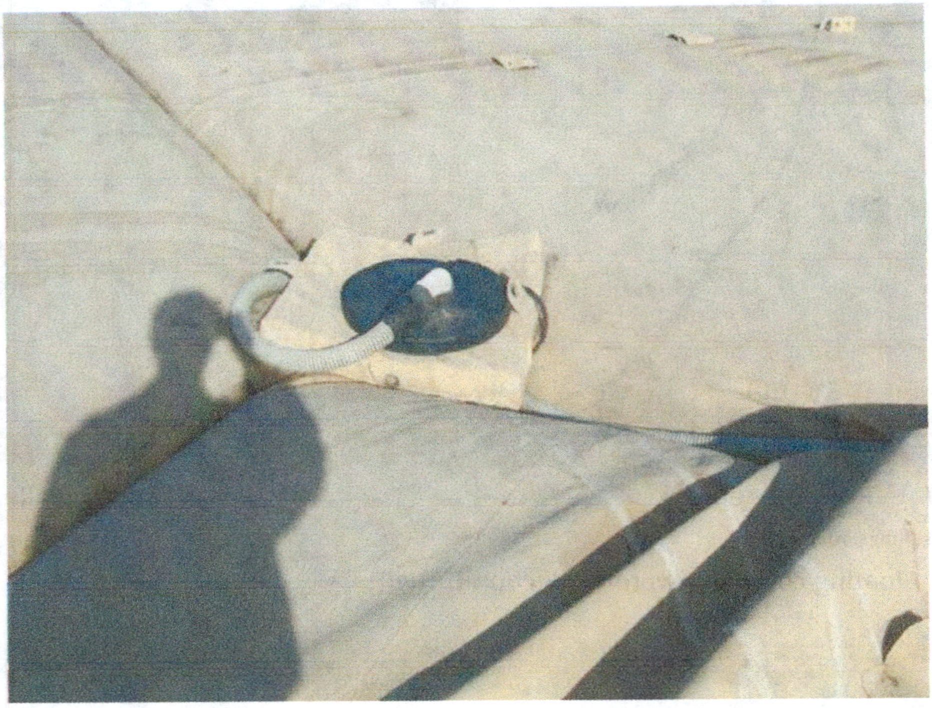

*Source:* Hilts Consulting Group

**Figure 2-13 Floating cover perforated can rainwater removal system**

out of service. Pumps should be equipped with automatic controls to enable the pumps to operate without the assistance of on-site personnel.

For gravity drain systems, no less than two gravity drains should be provided. The total number of gravity drains should be based on the surface water removal rate with one gravity drain out of service. A minimum of one gravity drain should be provided for each floating cell. The gravity drain hose should be suction hose construction with smooth interior surface, not less than 2-in. diameter, and reinforced to provide a factor of safety of 1.50 against collapse from external pressure or vacuum. Total hose length should be the horizontal run plus 1.33 times the maximum water depth. The outside of the hose should be ballasted to prevent hose flotation. The gravity drain should have the capability of being flushed. The top and bottom connection should have swivel joints to prevent hose distortion and kinking.

Surface water enhancement weights should be provided. These may be small-diameter sand-filled tubes or other appropriate mechanisms that may be laid loose on top of the floating cover after the reservoir is placed in service. The location is determined in the field since the purpose of these tubes is to direct localized ponded surface water to the surface water removal system. The project specifications should specify the number of surface water enhancement weights to be provided based on the surface area of the floating cover and the complexity of the design. A general rule of four surface water enhancement tubes per acre should be considered.

**Inflation.** Floating covers can be fitted with reasonably convenient and safe means of inflation and deflation to provide personnel access underneath while the floating cover is inflated (Figure 2-14). Some of these features include inflation hatches, blowers, access hatches, access structures, permanent or temporary floating cover weights, and in-reservoir slope stair steps.

*Source:* Hilts Consulting Group

**Figure 2-14 Floating cover inflation for inspection (reservoir empty)**

Although a floating cover may be fitted with all possible state-of-the-art inflation provisions and controls, inflation of the floating cover still must be attended by trained personnel 100 percent of the time during inflation for two important reasons.

- The first is to guard against over-inflation, which could lead to rupture of the floating cover.
- The second is to implement contingency plans to rapidly deflate the floating cover in the event of high winds. While inflated, the floating cover should be kept in as low a profile as possible.

**Repair floats.** Repair floats are recommended under the floating cover to permit future repairs to the floating cover material. A minimum of one repair float per floating cell should be provided, preinstalled, and remain under the floating cover for immediate use. Alternatively, repair floats may be deployed through hatches while the reservoir is in service.

These movable repair floats serve two functions.

- In the event of a hole or tear in the floating cover, the repair floats should be immediately moved to a location directly below the damage to elevate it above the surface level of the floating cover geomembrane until a permanent repair can be made. This precaution will prevent contamination of the contained water by any surface water.
- The repair floats facilitate the permanent repair while the floating cover is in service by holding the floating cover geomembrane above the contained water, so the affected area is dry.

Repair floats are usually made of closed-cell flotation foam of the same type used on the floating cover, and they are fully encased in the same geomembrane material as the floating cover. Repair floats should be round or octagonal in shape to facilitate movement under the floating cover. The ends of the repair floats should be rounded or chamfered. Repair floats can be moved from one location to another under the floating cover by a worker simply "walking" them into position; the repair floats roll or slide under the floating cover geomembrane while the worker walks behind them.

## Piping

Separate dedicated inlet and outlet facilities should be provided, rather than a single combined inlet/outlet. The separate inlet and outlet facilities are desirable for improved water quality and should be adequately spaced to promote circulation and minimize stagnant water storage zones.

**Inlet piping.** Inlet piping should be designed to avoid high-velocity flows and impact of water directly on the floating cover. Baffles or other means may be considered to direct incoming water away from the floating cover. Inlet configuration may be enlarged to reduce water entrance velocity.

**Outlet piping.** Outlet piping and/or the floating-cover system should be designed with a vacuum break or other means to prevent the floating cover material from being drawn into outlet pipes at low water levels. Outlets should ideally be flush with the reservoir floor or slopes to minimize excess floating cover material and potential for ponding surface water. The outlet configuration may be enlarged to reduce water exit velocity and reduce the potential for the floating cover to enter the outlet pipe. Outlet facilities may include silt stops, bar screens, pipe grillages, or bar grating.

**Overflow system.** The overflow system for a floating cover should be designed to prevent the water level from rising above the maximum design water depth and to prevent wind from causing inflation of the floating cover. The design should include provisions to prevent the floating cover from being drawn into the overflow system. Additionally, design should prevent the intrusion of small animals and rodents.

**Washdown water.** System design should provide water for cleaning purposes. Such provisions could be in the form of a piping system and fire hydrants near the reservoir perimeter.

## Material Selection

Refer to ANSI/AWWA D130 for floating cover geomembrane material requirements. See NSF/ANSI/CAN 61, Drinking Water System Components—Health Effects, for additional requirements for materials in contact with potable water. Additional state and local requirements may be applicable.

**Exposure.** Floating cover geomembrane material must be acceptable for potable water storage to all agencies having jurisdiction. The geomembrane material must be resistant to ultraviolet light, ozone, effects of continuous outdoor weathering, and contact with all common chemicals used to maintain water quality. The geomembrane material must continue to maintain its flexibility over the range of ambient temperature conditions at the installation site.

Physical properties of floating cover materials should include the optimum combination of tensile strength, elongation, tear strength, puncture resistance, and stress-cracking resistance to meet the special requirements of a floating cover.

All metals that are in direct contact with the potable water or water vapor should be either of corrosion-resistant material or coated with an NSF-certified coating. Those metals that are above the maximum water level and not exposed to reservoir water or water vapor may be coated or hot-dipped galvanized at the discretion of the engineer.

**Stress resistance.** Floating cover geomembrane material should be resistant to the fatigue of continuous flexing and the effects of wind, puncture, debris, and other stresses that could be encountered during the life of the floating cover. The floating cover geomembrane material must allow for easy and reliable seaming in the field, both during installation and for repair work thereafter.

**Construction.** Floating cover geomembrane material should be 45-mil nominal thickness, minimum 3-ply construction, and fabric reinforced. The seam shear strength should be not less than 90 percent of the parent material tensile strength.

**Life expectancy.** Past performance of floating cover geomembrane materials should be evaluated and a conservative economic life expectancy assigned. The manufacturer's and fabricator's warranties, if any, should be analyzed and considered to arrive at an anticipated useful life. A geomembrane floating cover 20-year service life may be obtained when:

- the material is properly manufactured and fabricated,
- an adequate design is provided,
- the floating cover is properly installed, and
- an adequate operations and maintenance plan is developed and implemented.

## Seams

Floating cover geomembrane material is fabricated in long strips called rollstock. For construction of a floating cover, the rollstock is factory-seamed together into large panels by a fabricator. The panels are then shipped to the site and field-seamed by the installer to form the final product. Depending upon the type of geomembrane material, seams can be made with heat, adhesives, or extruded parent material.

**Seam strength.** Seams should be of standard construction and compatible with the type of geomembrane material specified. Seam shear strength should be a minimum 90 percent of the parent material tensile strength and should not fail in the plane of the seam. The minimum overlap distance from scrim-edge to scrim-edge should be specified. All factory- and field-seam bonding should extend to the edge of the sheet so that no loose edge appears on the top side of the sheet. Minimum seam peel strength should be specified in the project specifications and should not fail in the plane of the seam and not fail along the weld interface.

**Seaming procedure.** The seaming procedure should be specified in detail, and it must not conflict with the manufacturer's recommendations. Specifications should provide for suspension of work in the event of adverse weather conditions at the project site that could affect seaming operations. Particular effort should be expended to prevent windblown dust or other debris from contaminating seams during preparation in the field.

**Adhesives or solvents.** Seams requiring adhesives or solvents should employ products recommended by the geomembrane material manufacturer. Such products must be approved for potable water. The products should contain no extractable ingredients that, when dissolved in water, would exceed applicable drinking water standards. Adhesives and bonding solvents should be checked for compatibility with manufacturer's recommendations and acceptance by the appropriate regulatory agencies.

**Testing and inspection.** Without exception, all seams made in both the factory and field must be tested for strength. Strength testing should be performed with physical samples according to ANSI/AWWA D130. Watertightness testing may be performed with a vacuum test chamber method. Inspection of all seams should include both a visual examination and an air-lance test for all factory and field seams. Air-lance testing identifies a defective seam by lifting the leading edge and creating a change in sound as the air is blown across the edge of the seam, usually causing the loose edge of the material to flutter. Only experienced personnel should perform seam testing and inspections.

During shop fabrication and field installation, trial seam installations should be performed, tested, and verified as passing before commencing with seaming production. Trial seam samples should be tested for shear and peel strengths and satisfy the minimum strength requirements stated in the project specifications. Trial seam testing should be performed at the beginning of each shift, near the midpoint of each shift, and upon resumption of work following any extended breaks during the shift. Seam samples should also be tested following any changes in seaming procedures, seaming equipment settings, or environmental conditions that affect seamability. Seam samples are used to verify seaming equipment settings (temperature, speed, etc.) relative to the daily weather conditions (temperature, humidity, winds, etc.) that can affect the overall seam strength. Trial seams should be made under the same conditions as production seams will be made.

Quality control destructive seam samples should be completed in accordance with ANSI/AWWA D130.

## Quality Assurance/Quality Control

Appropriate quality assurance/quality control plans should be developed for the material manufacturing, shop fabrication, and field installation. Strict adherence to these plans should be required and may be verified by the owner or representative inspector. The fabricator should certify that factory-fabricated panels are produced in accordance with the geomembrane material manufacturer's requirements and project specifications for seaming and patching. The fabricator should be responsible for factory seams, patches, and splices. The installer should be acceptable to the geomembrane material manufacturer and should be responsible for all field seams, patches, and splices. The contractor should provide an overall warranty on workmanship, factory and field seams, and all aspects of the floating cover installation. At the completion of construction, the contractor should certify that the floating cover was installed in accordance with ANSI/AWWA D130 and the project specifications. All quality assurance test reports and documents for the manufacturing, fabrication, and installation should be submitted to the owner for record purposes.

The engineer shall coordinate the project inspection requirements with the owner. An inspection program shall be developed that may include any or all of the following: manufacturing inspections, fabrication inspections, and field installation inspections. Inspections shall be performed by a qualified inspection agency, the engineer, or a combination. Inspection activities may include:

- certification of seaming equipment settings;
- observation of manufacturer's, fabricator's, and installer's quality assurance/quality control plan;
- third-party material conformance testing; and
- third-party material and seam destructive testing.

The project specifications shall provide the necessary notification requirements and delineate the inspection and third-party testing responsibilities between the owner and contractor.

## Extractables

Any adhesives, chemical fusion agents, or organic solvents or cleaners used in the fabrication, seaming, or installation of a floating cover should follow recommendations of the geomembrane material manufacturer. The materials should include no extractable ingredients that, when dissolved in water, would exceed applicable drinking water standards established by the appropriate regulatory agencies. Materials must be approved for contact with potable water.

### Warranty

The geomembrane manufacturer should warrant that the floating cover material is suitable for the intended use as a tensioned floating cover for a potable water reservoir, acknowledge that the material will be subjected to long-term exposure, and certify its compliance with the requirements of ANSI/AWWA D130. The geomembrane manufacturer should also warrant the floating cover material for the specified number of years stated in the engineer's project specifications. The geomembrane material manufacturer should warrant that all geomembrane materials are in compliance with the project specifications.

The fabricator should warrant that all factory seams are in accordance with the project specifications and that seaming procedures are performed in accordance with the geomembrane material manufacturer's recommendations. The warranty should affirm that the geomembrane material is free from workmanship defects at the time of delivery to the job site and warrant against fabrication workmanship defects for a period stated in the engineer's project specifications.

The installer should warrant that all field seams and the overall floating cover installation are in accordance with the project specifications and that the installation and field-seaming procedures are performed in accordance with the geomembrane material manufacturer's recommendations. The warranty should affirm that the geomembrane is free from workmanship defects and warrant against fabrication workmanship defects for the period stated in the engineer's project specifications.

The geomembrane material manufacturer, fabricator, and installer should certify that all seaming methods, cleaners, adhesives, solvents, and any other components used in the manufacturing, fabrication, and seaming process are safe and suitable for use in potable water reservoirs and are acceptable to the appropriate regulatory agencies.

## FLOATING COVER INSTALLATION

### Qualifications

The qualifications and experience of the manufacturer, fabricator, and installer should be verified by the owner before awarding the contract to supply a floating cover.

### Geomembrane Manufacturer's Experience

The geomembrane manufacturer should provide evidence of suitable qualifications and experience to manufacture the geomembrane materials called for in the project specifications.

### Certification for Potable Water Use

The geomembrane material manufacturer should submit certification to the owner that the proposed geomembrane is acceptable to the appropriate regulatory agencies for use in potable water facilities.

### Shop Drawings

Before starting fabrication, the installer should submit shop drawings as called for in the project specifications. The drawings should indicate the job number, description, and location. Shop drawings should show field-seam locations; top anchor details; details of floats and access hatches; thickness, color, name, and technical description of the geomembrane material; and factory- and field-seaming details. Reviewed and approved shop drawings should be returned to the installer by the owner.

Details of field and factory seams should be described in the shop drawings through use of diagrams and written descriptions. Seam overlaps, tolerances, and methods of seaming are important to the geomembrane material manufacturer's warranty and should not be changed by the fabricator or installer without the manufacturer's approval.

## Protection of Materials

After factory fabrication, all geomembrane panels should be packaged and fully enclosed to prevent damage during shipment, storage, and job site positioning and operations. Pallets or rolls should not be stacked. Packaged, factory-fabricated panels should be stored until needed in their original, unopened packages off the ground in a dry area completely shielded from the sun's direct rays. If stored outdoors, panels should be protected under a light-colored, heat-reflective opaque tarp in a manner that provides a 6-in. (150-mm) minimum free-flowing air space between the package and the tarp.

## Installation

The installation of the floating cover should proceed according to the approved shop drawing layout for field seams and details. Particular effort should be expended to prevent windblown dust or other debris from contaminating field seams during seaming operations. A qualified inspector should closely monitor installation for compliance with the approved design and specification documents. Figures 2-15 to 2-17 illustrate field seaming for a polypropylene membrane.

*Source:* Hilts Consulting Group

**Figure 2-15 Wedge welder seaming**

*Source:* Hilts Consulting Group

**Figure 2-16 Hot air welder seaming**

*Source:* Hilts Consulting Group

**Figure 2-17 Hot air hand seaming**

This page intentionally blank.

Chapter 3

# Operation, Inspection, and Maintenance Guidelines for Geomembrane Floating Covers

## GENERAL

Water industry experience has clearly established the critical need for a well-defined, regularly performed site-specific operation, inspection, and maintenance program with written documented procedures. Such a program is essential for successful long-term performance of a geomembrane floating cover. Without this well-defined operation, inspection, and maintenance program and its necessary documentation, all geomembrane floating cover designs will ultimately develop a potential for contamination of the stored water and failure. Reservoir geomembrane floating covers, regardless of design, are no different from any other major component of a water treatment or storage facility; they require regular inspection, cleaning, and repair with thorough documentation.

This chapter lays out guidelines to assist water system operators in the operation, inspection, and maintenance of geomembrane floating covers used in potable water storage reservoirs. The scope of these guidelines is limited to procedures that have been found to work successfully at various water utilities in the United States. Special emphasis is placed on procedures adopted by utilities that operate in cold climates.

Operating experiences with geomembrane floating cover systems began during the mid-1960s, with the first successful installation in 1965 at Chino, Calif., on a 1.5-mil-gal (5.7-ML) finished water reservoir. Some of the first geomembrane floating covers used in cold regions were installed at Pottstown, Pa., in 1968 and Manchester, N.H., in 1970.

Many early installations of butyl or ethylene propylene diene monomer (EPDM) floating covers began in later years to exhibit a reduced tolerance in field seams because of the bending and twisting action caused by ice movement and reservoir level fluctuations. Elastomeric geomembrane materials, such as chlorosulfonated polyethylene (CSPE-R), have been used for floating covers since the early 1970s; since the mid-1990s, polypropylene (PP-R) and Elvaloy™ have also been used. These materials are considered the industry standard for potable water applications, with many millions of square feet of elastomeric geomembrane material installed over recent decades.

# OPERATION

## Reservoir Startup

After the installation is complete and most residual solvent used in the seaming process evaporates, the filling of the reservoir may begin. At low operating levels, the floating cover is susceptible to damage by water entering the reservoir at high flows. A reservoir-specific fill plan should be developed outlining flow rates, durations, reservoir water levels, and total volume of water. The fill plan should be coordinated with the reservoir disinfection plan. The initial fill should commence at reduced flows until the potential for damage from impinging water onto the floating cover is greatly reduced. The quantity of air introduced underneath the floating cover should be reduced whenever possible. Entrapped air should be "walked" to the nearest venting mechanism. The floating cover should be closely observed during the filling period. The floating cover should be inspected several times throughout the filling period for leaks in the floating cover geomembrane material and at special locations (e.g., surface water collection troughs, inlets, grillages). If leaks are discovered in the surface water collection trough, the filling should immediately stop, and all leaks should be repaired. Upon completion of filling, the floating cover should again be thoroughly inspected for leaks. All leaks should be repaired in accordance with an approved operations and maintenance plan.

The owner should anticipate additional water quality sampling during the filling and startup period.

## Disinfection

Disinfection of the reservoir should be performed by the owner. The reservoir should be disinfected after the completion of the initial floating cover installation and after removing the reservoir from service for cleaning or repairs. The most common disinfection method for floating cover installations is ANSI/AWWA C652, Disinfection of Water-Storage Facilities, Method 3. The suitability of the disinfection method and concentration of chemicals should be confirmed by the geomembrane material manufacturer to prevent potential damage to the floating cover geomembrane material and other system components. The main disinfection objective should be to provide a free residual chlorine level of 2 mg/L after 24 h contact time and a bacteria-free condition in the entire reservoir volume before returning it to service. The required amount of disinfectant should be added to the reservoir at low water level through hatches or preferably introduced into the incoming water lines.

The disinfection procedure should be submitted to and approved by the applicable state regulatory agencies. Water quality tests for bacteria, organics, and inorganics should be performed before returning the reservoir to service.

## Cold-Weather Considerations

Water system operators should closely monitor and, if possible, control reservoir operating levels whenever ice has formed on or beneath the surface of a floating cover. Wide fluctuations in reservoir levels can result in ice damage to a floating cover, particularly along sloping sidewalls. A reservoir operating range not exceeding 1 vertical ft (0.3 m) is recommended during freezing weather conditions.

## Reservoir Dewatering

When a reservoir is dewatered, the floating cover shall lie directly on the reservoir floor and side slopes. As with the filling of the reservoir, the dewatering should also be performed in accordance with a comprehensive dewatering plan. The dewatering rate should take into account those outlet structures that have the potential to draw the floating cover into the outlet during rapid dewatering operations. As the water surface approaches the outlet structures, the dewatering rate should be reduced to minimize the potential for damage to the floating cover.

## Surface Water Removal

Surface water that accumulates on the floating cover from snowmelt and rainfall should be removed as soon as possible. The removal of surface water should be performed in a prompt manner by the surface water removal system. Complete removal of surface water immediately before the onset of freezing weather is considered essential to successful floating cover operation. Removal of this water minimizes ice formation on top of the floating cover, thus minimizing damage to the geomembrane material and possible contamination of the potable water. Submersible electric sump pumps, operated by manual or automatic timing devices, have been successfully used to remove surface water (Figures 2-11 to 2-13).

## Safety

Proper safety procedures should be observed at all times when inspection, operation, or maintenance work is being performed with the floating cover in service. Wet surfaces of floating cover materials are extremely slippery. Personnel should exercise caution when walking on reservoir side slopes. Caution should be used when near surface water collection troughs, as these troughs may be over 6 ft deep.

The following minimum safety rules should be adopted and enforced by the owner:

- No fewer than two workers should be present at a location when work of any nature (inspection or maintenance) is taking place on a reservoir floating cover. Workers shall maintain visual contact at all times.
- A life jacket or flotation vest should be worn by any individual working on a floating cover. An adequate supply should always be available on site.
- All personnel should have a sharp knife for the purpose of cutting oneself free if accidentally trapped under the floating cover. Knives should be properly sheathed and secured to prevent personal injury and damage to the floating cover.
- A convenient means of communication should be available (e.g., two-way radios).

- A properly fitting, insulated wet suit should be worn when working on a floating cover during cold weather.
- A minimum of 100 ft (30 m) of rope attached to a suitable life ring should be readily available to workers while work is taking place on the floating cover.
- Stairs, textured walkways, knotted ropes, or rope ladders should be provided on reservoirs with steep side slopes to aid in safely entering and exiting the reservoir.
- Electrical tools used to facilitate repairs to the floating cover while it remains in service *must* be protected by an appropriate ground fault-interrupting (GFI) device at the source of power.
- Utility knives to perform all work shall be hooked blades.

Copies of all MSDS sheets for all chemicals used in association with a floating cover should be posted on site.

## INSPECTION

Complete inspections of the floating cover should be performed at regular intervals to check for possible leaks, buildup of debris, separation of seams, damage caused by vandalism, and accumulation of surface water on the floating cover. Inspections should also check for proper operation of pumps and drains. A careful record of any required maintenance should be developed and preserved for future reference.

A site-specific operations and maintenance plan should be developed for each floating cover facility. The following factors should be considered when developing site-specific procedures:

- climatic conditions,
- location of floating cover,
- age and condition of floating cover,
- type of floating cover material,
- reservoir size,
- jurisdictional and regulatory agency requirements, and
- other factors.

The operations and maintenance plan should be submitted to the applicable regulatory agencies for review and approval. An inspection checklist should be developed listing all necessary observations to be made during each inspection. Inspection records should be maintained for historical purposes and as a means of demonstrating compliance with any regulatory requirements.

### Visual Inspection (Off Floating Cover)

Operator performs a regular perimeter inspection (off the floating cover) to visually detect any problems. It is recommended that the inspection be completed daily, but intervals may vary depending on the site-specific installation. Inspection records should be kept, including the date and time of inspection, a description of findings, and sketches to identify areas chosen for closer inspection or repair.

### Maintenance Inspection (On Floating Cover)

A detailed inspection on the floating cover should be performed on a monthly basis, at a minimum. Floating cover maintenance inspections should include the following procedures:

- Traverse the floating cover at a maximum of 25-ft (7.6-m) intervals, ensuring all shop and field seams are inspected.
- Check all hatches, vents, and surface water collection areas along with the overall surface of the floating cover for accumulation of debris, surface water, and signs of leakage.
- Remove any debris by manually sweeping, vacuuming, or blowing air.
- Mark all holes and tears that are found. Patch damaged areas utilizing closed-cell repair floats to maintain a gap between the finished water and the floating cover area under repair.
- Adjust sand-filled surface water enhancement weights to direct any areas of standing water into surface water collection troughs.
- Operate surface water removal pumps to ensure that they are free of blockages and operating properly.
- Note all comments on the inspection checklist including type and location of all repairs made, equipment maintenance, and cleaning. Sign and date the checklist.

## Underwater Inspection

Diver and remotely operated vehicle (ROV) safety provisions must be observed in accordance with all local, state, and federal regulations. Underwater inspection by divers or ROVs are usually performed at least annually, or more frequently if necessary, to investigate concerns regarding damage to the floating cover or equipment. The owner may consider additional underwater inspections after heavy winters with the potential for or suspicion of experiencing damage from freezing conditions. The reservoir should ideally be in the near-full condition for underwater inspections. Two people in addition to the diver must be present at all times as required by Occupational Safety and Health Administration (OSHA) regulation. All equipment, including dry suits, must be dedicated for potable water use only. The following procedures are to be used in these underwater inspections:

- Wash all equipment in 200 mg/L chlorine solution in accordance with ANSI/AWWA C652 requirements before entry into the reservoir.
- Inspect and document any gravity drain hoses from top to bottom and the attachment to the surface water collection troughs.
- Inspect and document surface water collection troughs.
- Inspect and document floating cover conditions at reservoir inlets and outlets, valves and gates, grillages and floating cover support structures, and surface water collection troughs.
- Compare documentation taken to previously recorded video and/or photos on file.
- Prepare a written report detailing findings, including video and/or photos, with specific maintenance recommendations.

All underwater inspections shall be videotaped and/or photographed. The sequence of inspection shall be scripted by the owner and followed for every periodic underwater inspection to provide a constant and uniform comparison with previous inspections.

*Source:* Hilts Consulting Group

**Figure 3-1 Floating cover cleaning**

## MAINTENANCE

### Cleaning

Floating cover surfaces should be thoroughly cleaned by removing dirt, leaves, and other foreign matter (Figure 3-1). Any foreign objects (such as rocks and bottles) on the floating cover should be removed as soon as possible to prevent buildup of debris and possible damage to the geomembrane material.

Organic materials, such as leaves and bird droppings, should be removed from the floating cover, particularly from sumps where surface water accumulates. Organic matter, silt, and other fine debris should be regularly removed by sweeping, blowing, or vacuuming the floating cover surface and/or flushing with water. In the event of a leak in the floating cover geomembrane material, the stored water could be contaminated by any organic material on top of the floating cover surface. Prompt repair of leaks is essential.

Only rubber-soled shoes should be worn by personnel performing inspection or maintenance work on the floating cover. The reservoir should be equipped with a suitable central sump or drain to convey surface water and suspended solids to waste.

### Repair Floats

If inspection reveals a hole or tear, a repair float should be immediately moved to a location directly below the damage to elevate it above the surface level of the geomembrane floating cover until a permanent repair to the hole or tear can be made.

## Geomembrane Repairs

Floating cover repairs should be performed only by properly trained personnel using materials and methods outlined in the operations and maintenance manual. Repair work should not be attempted with the floating cover in service if repaired areas can come into contact with the potable water during the procedure. Because taste- and odor-causing chemicals may be used in making the repairs, extreme caution must be taken to ensure that any excess chemicals are thoroughly dissipated once repairs are completed.

If a repair float is not already in place, a repair to a horizontal surface of the floating cover can be performed by inserting a float through an access hatch and then positioning the repair float beneath the damaged area to raise the floating cover surface above the water line. Repairs along reservoir side slopes can often be completed with the floating cover in service by temporarily lowering the water level in the reservoir.

This page intentionally blank.

Chapter 4

# Design and Installation of Geomembrane Liners

## LINER DESIGN

Using technology developed during World War II, the first geomembrane liners were introduced in the early 1950s. Thermoplastics, such as polyvinyl chloride (PVC) and polyethylene (PE), emerged, followed by cured synthetic rubbers (butyl and ethylenepropylene-diene monomer [EPDM]). The late 1960s brought development of the first thermoplastic elastomer, chlorosulfonated polyethylene (CSPE-R), known as Hypalon™. For the first time, this material combined the reliable seaming properties of thermoplastics with the superior weathering and ultraviolet resistance of cured rubbers. In recent years, numerous advancements have improved polymer technology as well as manufacturing equipment and processes with reinforced geomembranes. Improved geomembrane liner constructions of reinforced CSPE-R and polypropylene (PP-R) have been used throughout the United States. Each of these geomembrane materials has different physical characteristics and methods of construction and repair that should be thoroughly evaluated before a geomembrane product is specified.

### General Areas of Concern

This section reviews areas of concern that the purchaser must recognize in several categories.

**Engineer qualifications.** The drawings and specifications for a geomembrane liner system should be prepared by a licensed engineer having specialized knowledge and experience in the design and operation of flexible geomembrane liner systems. Design drawings should include:

- liner surface area,
- overflow elevation,
- high operating level,
- normal operating range,
- low operating level,
- reservoir capacity,
- inlet and outlet flow capacities,
- geomembrane liner material (type, thickness, color, reinforcing, etc.),
- seaming procedures,
- other miscellaneous support material(s),
- underdrain system,
- leakage collection and monitoring system,
- attachment to structures,
- air vents,
- perimeter attachment, and
- penetrations.

Design calculations should, at a minimum, address maximum system design level leakage, actual typical daily performance level leakage, and liner stress.

**Degree of reliability desired.** If the reservoir is a critical component of the water system, extensive out-of-service time for repairs and maintenance may not be acceptable. In such a case, a liner material with a proven record of satisfactory performance should be selected. In addition, the reservoir could be designed as a dual basin, so that one half can be taken out of service independently while the other half remains in use. Alternatively, a bypass system can be provided to facilitate the reservoir being taken out of service.

**Economics (budget limitations).** Generally, geomembrane liners are the most economical alternatives for reservoir liner systems based on initial construction cost. However, a life-cycle total cost comparison should consider the periodic replacement and maintenance expenses. Maintenance expenses should be based on historic costs determined under similar conditions of use.

**Operations and maintenance methods and equipment.** Once the geomembrane liner is installed, only foot traffic with smooth-soled rubber shoes or boots should be allowed on its surfaces. Internal combustion equipment should not be operated inside the reservoir for maintenance or repairs.

**Jurisdiction and/or regulatory agency requirements for types of liner materials.** The geomembrane liner material, chemical fusion agents, adhesives, and similar items should be suitable for contact with potable water. Materials should not contain any ingredients that, when dissolved in water, would exceed established applicable water quality standards or add objectionable taste or odor. The geomembrane liner material should be verified as acceptable to appropriate regulatory agencies and in compliance with the requirements of ANSI/AWWA D130.

**Equipment.** Equipment should not be operated on top of an installed geomembrane liner unless absolutely necessary. Such equipment, if used, should have smooth rubber tires, and the foundation bearing strength should be sufficient to support the equipment's wheel load without deformation. Equipment should not operate on slopes or where sliding may occur.

**Foundations.** Generally, a geomembrane liner must be installed over a stable and properly prepared substrate or foundation. A geomembrane liner should be used as a watertight barrier and should not be designed as a load-bearing system. When provisions are made in the design, a geomembrane liner can accommodate minor differential settling because the material will conform to the contours of the settled foundation. The foundation should be designed by an engineering firm experienced in the design of foundations for geomembrane liners.

## Environmental Considerations

**Site conditions.** The geomembrane liner design should account for groundwater conditions that may be encountered at the site during the in-service life. Ideally, the reservoir bottom should be located above the historically high groundwater elevation.

Reservoir side slopes should ideally be no steeper than 2 horizontal units to 1 vertical unit. Reservoirs with steeper side slopes should have special provisions to increase worker safety during construction and required maintenance and operation activities. Internal reservoir surfaces should be smooth, flat, and uniform. Reservoir side slopes should be straight and uniform.

Other site considerations include bypass piping for the ability to isolate the reservoir without disrupting water delivery, and clearances from sewage facilities. The engineer should consult applicable state waterworks standards for regulatory requirements.

**Climatic conditions.** It is generally recognized that in cold-weather climates subject to ice load, wide fluctuations in reservoir water levels can result in damage to a geomembrane liner unless special precautions are taken. An air bubbler or other system should be considered to prevent ice formation around gate structures and other penetrations. Design should prevent ice abrasion on geomembrane material along the side slopes as the water surface level is raised and lowered, since this abrasion can cause geomembrane liner material to split, seam failures, and anchorage failures. To avoid excessive wear on the geomembrane liner, water-level fluctuations should be minimized during winter. Installation of a membrane chafing strip may be warranted to resist abrasion along reservoir sidewalls.

**Wind.** An exposed geomembrane liner will be subjected to wind uplift. Provisions must be included in the design to counteract the forces exerted on the exposed liner from wind uplift. Local wind speed should be determined from historic records, local building officials, or other rational and comparable method.

**Construction sequence.** In a reservoir with a rigid roof system, the geomembrane liner should be installed after roof installation has been completed to minimize the potential for damage during subsequent construction. This procedure can result in an enclosed reservoir structure. When bonding adhesives or solvents (chemical fusion solutions) are being used to create field seams in an enclosed reservoir, the reservoir should be properly ventilated to avoid accumulation of solvent fumes.

## Design Considerations

**Leakage.** With current state-of-the-art geomembrane design, inspection, and construction techniques, leakage from a reservoir with a geomembrane liner should be within the normally expected limits for reservoirs of alternative materials. Leakage should be monitored by measuring the water level at regular intervals and comparing data after compensating for rainfall, evaporation losses, and ensuring that no leakage occurs through valves and piping. For single-liner systems, measuring leakage from underdrains alone does not provide an accurate measurement because some leakage may percolate into the subgrade before reaching the underdrain, and groundwater may enter the underdrains, thereby distorting leakage measurements. Seals around penetrations, terminations of geomembrane liners at walls, piping sumps, and seams are potential leakage areas.

The maximum allowable leakage rate should be established by the engineer and included in the bid documents in accordance with ANSI/AWWA D130. The geomembrane liner specifications should include a maximum allowable leakage rate based on rational engineering principles. The maximum allowable liner-leakage rate shall account for all variables, including but not limited to:

- liner area,
- frequency of material defects,
- size of material defects,
- reservoir head,
- hydraulic conductivity of materials over and under the liner material,
- placement and type of materials installed on top of geomembrane liner,
- number of underwater seals to structures, and
- overall complexity of installation.

The frequency of defects may range from 1–5 holes per acre, depending on a rigid quality assurance/quality control plan, manufacturing inspection, fabrication inspection, and full-time on-site inspection. In addition to leakage from material defects, the leakage rate shall include permeation of water through the geomembrane liner material.

For large reservoirs with double-liner systems, the design should consider subdividing the liner-leakage collection system into smaller cells with individual liner-leakage collection pipes. This will assist in the monitoring and location of excess leakage. The liner-leakage collection system shall be designed for the maximum allowable leakage rate multiplied by a factor of safety of 10 to account for catastrophic liner failures and long-term clogging. Allowable liner-leakage rates for each subdivided cell shall be based on the same variables listed previously.

**Drainage blanket.** A drainage blanket or leak collection layer should be provided immediately beneath the geomembrane liner. The drainage blanket consists of a highly permeable material (e.g., drain rock or geosynthetic drainage net). The drainage blanket serves to collect any liner leakage and to relieve pressure on the backside of the liner from high groundwater conditions. The transmissivity of the permeable material should be greater than the design level flow rate (leakage and groundwater) to prevent standing water on the backside of the liner. A drainage blanket from drain rock usually consists of a rounded granular material graded such that 100 percent passes through a ⅜-in. (9.6-mm) sieve. If the geomembrane liner is to be supported directly on top of the drainage layer, provisions should be made to protect the geomembrane against puncture due to the roughness of the drainage material. Use of a geotextile fabric to protect the liner should be considered. A geosynthetic drainage blanket net may be a composite drainage course (single- or double-sided) or a drainage net with separately laid geotextile layer(s) at locations determined by the engineer.

To promote free drainage, the reservoir floor and drainage blanket should be sloped toward a low point at a minimum grade of ½ percent.

**Underdrains.** Underdrains are recommended below all geomembrane liners. The underdrains serve to collect leakage and groundwater from the drainage blanket layer. Underdrains typically consist of perforated pipes in lined trenches backfilled with permeable drain rock. Underdrains are typically a grid of buried pipes in the reservoir floor that manifold together and terminate at a single location outside of the reservoir perimeter where flow may be conveniently monitored.

**Geomembrane liner subgrade.** The geomembrane liner should be continuously supported by a flat, smooth subgrade surface that will retain its stability under all moisture conditions. The reservoir subgrade surface should be smooth without sharp edges. Widths of cracks or holes should be limited to a maximum unsupported width of 1⁄16-inch (1.6 mm) to prevent possible rupture under water pressure. Cracks with vertical displacement should be leveled. Caution should be exercised for substrates of newly applied asphalt or similar materials containing volatile substances that could chemically attack the geomembrane liner material.

**Penetrations.** Particular attention should be given to the method of sealing and anchoring the geomembrane liner around pipes and columns penetrating the geomembrane liner. Special fittings, such as reglets, embedded items, and clamps, can be used to achieve watertight seals. Prefabricated vacuum-formed boots should be considered. Penetrations should not be bundled together.

**Perimeter anchorage.** A concrete anchor curb or anchor trench backfilled with compacted soil or concrete/slurry should be provided for perimeter anchorage. The point of anchorage (curb batten bars or trench crest) shall be a minimum of 24 in. above the water surface elevation at overflow condition. The engineer should verify with regulatory agencies if additional freeboard is required.

**Piping.** The type of reservoir floating cover (if any) affects the geomembrane liner design provisions for inlet, outlet, drain, and overflow piping inside the reservoir. Inlet and outlet piping should be designed to avoid high velocities and impact of water directly on the geomembrane liner. Concrete splash pads or energy-dissipating structures should be provided to prevent damage to the geomembrane liner. The geomembrane liner should extend above the highest overflow water level plus freeboard.

Separate dedicated inlet and outlet facilities should be provided, rather than a single combined inlet/outlet. The separate inlet and outlet facilities are desirable for improved water quality and should be adequately spaced to promote circulation and minimize stagnant water storage zones.

A separate drain pipe at the reservoir floor low point is desirable to allow for cleaning by hosing the reservoir geomembrane liner during periodic maintenance. Provision should be made to provide water for cleaning purposes. Such provisions could include piping in the reservoir or a fire hydrant near the outside perimeter.

**Venting.** Generation or accumulation of gas or air under the geomembrane liner is a potentially serious problem. The drainage blanket and underdrain system must be vented and capable of relieving pressure from gas pockets under the geomembrane liner. If air or gas is allowed to accumulate under the geomembrane liner, the material rises off the subgrade—stretching the membrane and creating the potential of eventual rupture. Vents are provided at the high points, typically the reservoir perimeter, and spaced at regular intervals.

**Double liners.** Based on performance criteria, the engineer may consider a double-liner system. Double-liner systems are typically comprised of two geomembrane liner layers with a drainage media or blanket between the two layers. A double-liner system offers significant performance benefits:

- It provides two independent liner layers for redundancy.
- Leakage through the top layer is contained and isolated.
- Liner leakage cannot be mixed with groundwater and can be better quantified.
- The bottom layer offers additional protection from contaminated, nonpotable groundwater.

**Underwater location markers.** Underwater markers should be provided on top of the liner to allow accurate diver location during underwater inspections. A rational grid shall be established in two primary directions, with markers provided at regular spacing but not greater than 100 ft (30 m) in either direction.

## Material Selection

Refer to ANSI/AWWA D130 for geomembrane liner material requirements. See NSF/ANSI/CAN 61 for additional requirements for materials in contact with potable water. Additional state and local requirements may be applicable.

**Exposure.** Geomembrane liner material must be acceptable for potable water storage to all agencies having jurisdiction. (Refer to ANSI/AWWA D130 and NSF/ANSI/CAN 61). The material must be resistant to sunlight, ultraviolet light, ozone, effects of continuous outdoor weathering, and contact with all common chemicals used to maintain water quality. The material must continue to maintain its flexibility over the range of ambient temperature conditions at the installation site.

Physical properties of geomembrane liner materials should include the optimum combination of tensile strength, elongation, tear strength, puncture resistance, UV resistance, ozone resistance, and stress-cracking resistance to meet the special requirements of a geomembrane liner.

All metals that are in direct contact with the potable water or water vapor should be either of corrosion-resistant material or coated with an NSF-certified coating. Those metals that are above the maximum water level and not exposed to reservoir water or water vapor may be coated or hot-dipped galvanized, at the discretion of the engineer.

**Stress resistance.** Geomembrane liner material should remain flexible and be resistant to the effects of wind, puncture, debris, and other stresses that could be encountered during the life of the liner. The material must allow for easy and reliable seaming in the field, both during installation and for repair work thereafter.

**Construction.** Geomembrane liner material should meet the requirements of ANSI/AWWA D130. The seam shear strength should be not less than 90 percent of the parent material tensile strength.

**Life expectancy.** Past performance of a geomembrane liner material should be evaluated and a conservative economic life expectancy assigned. The manufacturer's and fabricator's warranties, if any, should be analyzed and considered to arrive at an anticipated useful life. A geomembrane liner service life of 20 years or more may be obtained when the material is properly manufactured and fabricated, an adequate design is provided, the geomembrane liner material is properly installed, and an adequate operations and maintenance plan is developed and implemented.

## Seams

Geomembrane liner material is fabricated in long strips called rollstock. For construction of a geomembrane liner, the rollstock is factory-seamed together into large panels by a fabricator. The panels are then shipped to the site and field-seamed by the installer to form the final product. Depending upon the type of geomembrane material, seams can be made with heat, adhesives, or extruded parent material.

**Seam strength.** Seams should be of a standard construction and compatible with the type of material specified. Seam shear strength should be a minimum 90 percent of the parent material tensile strength and should not fail in the plane of the seam. The reinforced geomembrane liner design should specify the minimum scrim-edge to scrim-edge overlap distance. All bonding for factory and field seams should extend to the edge of the sheet, leaving no loose edge on the top side of the sheet. Minimum seam peel strength should be specified in the project specifications, and the seam should not fail in the plane of the seam.

**Seam procedure.** The seaming procedure should be specified in detail, and it must not conflict with the geomembrane material manufacturer's recommendations. Specifications should provide for suspension of work in the event of adverse weather conditions at the project site that could affect seaming operations. Particular effort should be expended to prevent windblown dust or other debris from contaminating seams during preparation in the field.

**Adhesives or solvents.** Seams requiring adhesives or solvents should employ products recommended by the geomembrane material manufacturer. Such products must be approved for potable water. The products should contain no extractable ingredients that, when dissolved in water, would exceed applicable drinking water standards. Adhesives and bonding solvents should be checked for compatibility with the geomembrane material manufacturer's recommendations and acceptance by the appropriate regulatory agencies.

**Testing and inspection.** Without exception, all geomembrane material seams made in both the factory and field must be tested for strength. Strength testing should be performed with physical samples according to ANSI/AWWA D130. Watertightness testing may be performed with a vacuum test chamber or electronic leak detection methods. Inspection of all seams should include both a visual examination and an air-lance test for all factory and field seams. Air-lance testing identifies a defective seam by lifting the leading edge and creating a change in sound as the air is blown across the edge of the seam, usually causing the loose edge of the material to flutter. Only experienced personnel should perform seam testing and inspections.

During shop fabrication and field installation, trial seams should be constructed, tested, and verified as passing before commencing with seaming production. Seam samples should be tested for shear and peel strengths and satisfy the minimum strength requirements stated in the project specifications. Seam samples should be made and tested at the beginning of each shift, near the midpoint of each shift, and upon resumption of work following any extended breaks. Seam samples should also be tested following any changes in seaming procedures, seaming equipment settings, or environmental conditions that affect seamability. Seam samples are used to verify seaming equipment settings (e.g., temperature, speed) relative to the daily weather conditions (e.g., temperature, humidity), which can affect the overall seam strength.

Quality control destructive seam samples should be developed in accordance with ANSI/AWWA D130.

## Quality Control

Appropriate quality assurance/quality control plans should be developed for the geomembrane material manufacturing, shop fabrication, and field installation. Strict adherence to these plans should be required and may be verified by the owner or representative inspector. The fabricator should certify that factory-fabricated panels are produced in accordance with the geomembrane material manufacturer's requirements and project specifications for seaming and patching. The fabricator should be responsible for factory seams, patches, and splices. The installer should be acceptable to the geomembrane material manufacturer and fabricator and should provide a warranty on workmanship, field seams, and all other aspects of the geomembrane liner installation. At the completion of construction, the installer should certify that the geomembrane liner was installed in accordance with ANSI/AWWA D130 and the project's specifications. All quality assurance test reports and documents for the geomembrane material manufacturing, fabrication, and installation should be submitted to the owner for record purposes.

The engineer shall coordinate the project inspection requirements with the owner. An inspection program shall be developed which may include any or all of the following items: manufacturing inspections, fabrication inspections, and field installation inspections. Inspections shall be performed by a qualified inspection agency, the engineer, or a combination. Inspection activities may include:

- certification of seaming equipment settings;
- observation of manufacturer's, fabricator's, and installer's quality assurance/quality control plan;
- third-party material conformance testing; and
- third-party material and seam destructive testing.

The project specifications shall provide the necessary notification requirements and delineate the inspection and third-party testing responsibilities between the owner and contractor.

## Extractables

Any adhesives, chemical fusion agents, or organic solvents or cleaners used in the fabrication or installation of a liner should follow recommendations of the geomembrane material manufacturer. The materials should include no extractable ingredients that, when dissolved in water, would exceed applicable drinking water standards. Materials must be approved for contact with drinking water.

## Warranty

The geomembrane material manufacturer should warrant that the geomembrane liner material is suitable for the intended use as a geomembrane liner for a potable water reservoir; acknowledge that the material will be subjected to long-term exposure; and certify its compliance with the requirements of ANSI/AWWA D130. The geomembrane material manufacturer should also warrant the liner material for the specified number of years stated in the engineer's project specifications. The geomembrane material manufacturer should warrant that all geomembrane materials are in compliance with the project specifications.

The fabricator should warrant that all factory seams are in accordance with the project specifications and that seaming procedures are performed in accordance with the geomembrane material manufacturer's recommendations. The warranty should affirm that the geomembrane is free from workmanship defects at the time of delivery to the job site and warrant against fabrication workmanship defects for the period stated in the engineer's project specifications.

The installer should warrant all field seams and the overall geomembrane liner installation are in accordance with the project specifications and that the installation and field-seaming procedures are performed in accordance with the geomembrane material manufacturer's recommendations. The warranty should affirm that the geomembrane liner is free from workmanship defects for a period stated in the engineer's project specifications.

The geomembrane material manufacturer, fabricator, and installer should certify that all seaming methods, cleaners, adhesives, solvents, and any other components used in the manufacturing, fabrication, and seaming process are safe and suitable for use in potable water reservoirs and are acceptable to the appropriate regulatory agencies.

# LINER INSTALLATION

## Qualifications

The owner should verify the qualifications and experience of the geomembrane material manufacturer, fabricator, and installer before awarding a contract to supply a geomembrane liner.

## Geomembrane Manufacturer's Experience

The geomembrane material manufacturer should provide evidence of suitable qualifications to manufacture the geomembrane called for in the project specifications.

## Certifications for Potable Water Use

The geomembrane material manufacturer should submit certification to the purchaser that the proposed geomembrane is acceptable to appropriate regulatory agencies for use in potable water facilities.

## Shop Drawings

Before beginning fabrication, the installer should submit shop drawings as called for in the project specifications. The drawings should indicate the job number, description, and location. Shop drawings should show:

- field-seam locations;
- direction of shop seams;
- top anchor details;
- details of geomembrane liner seals at all penetrations (such as columns, structures, and pipes);
- geomembrane liner material thickness, color, name, and technical description of the geomembrane liner material; and
- factory- and field-seaming details.

Reviewed and approved shop drawings should be returned to the installer by the owner.

Field- and factory-seaming details, including overlays and tolerances, should be described in the shop drawings through use of diagrams and written descriptions. Seam overlaps, tolerances, and methods of seaming are important to the geomembrane material manufacturer's warranty and should not be changed by the fabricator or installer without the geomembrane material manufacturer's approval.

## Protection of Materials

After factory fabrication, geomembrane liner panels should be packaged and fully enclosed to prevent damage during shipment, storage, and job site positioning and operations. Pallets or rolls should not be stacked. Packaged, factory-fabricated panels should be stored until needed in their original, unopened packages, off the ground, in a dry area completely shielded from the sun's direct rays. If stored outdoors, panels should be protected under a light-colored, heat-reflective, opaque tarp in a manner that provides a 6-in. (150-mm) minimum free-flowing air space between the package or crate and the tarp.

## Stability of the Substrate

The installer should examine the substrate stability and identify any areas that may require further preparation. Before any geomembrane liner material is installed, the installer must determine that the substrate is stable and whether it is wet or dry. No geomembrane liner system will make a stable structure out of an inherently unstable substrate.

## Surface Texture of the Substrate

The installer should examine the texture of the substrate and determine whether it is suitable for installation of the geomembrane liner. The installer should not install geomembrane liner material on any substrate with a texture rough enough to cause damage to the liner. If surface preparation is necessary, it should be done before the geomembrane liner is installed.

## Installation

The installation of a geomembrane liner should proceed according to the layout for the individual panels as shown on the approved shop drawings. Shop-fabricated panels should be field-seamed in accordance with the approved shop drawings, and all penetration seals should be made as shown on the drawings. Particular care should be taken to keep windblown sand or dust from contaminating field seams during seaming operations.

## Leakage Testing

The reservoir should be filled at a reasonable rate (i.e., approximately 2 vertical ft [0.6 m] per day) continuing until the reservoir is full, unless a specific filling procedure is called for in the specifications. Leakage should be monitored by measuring the water level at intervals after compensating for rainfall, evaporation losses, groundwater entering underdrains, and ensuring that no leakage is occurring through valves and piping. Measuring leakage from output of underdrains alone does not provide an accurate measurement because some water may percolate to the subgrade before reaching the underdrain. Seals around penetrations, as well as terminations of the geomembrane liner at walls, piping sumps, and seams, are potential leakage areas.

While various techniques are used to locate leaks in liners in service, the possibility of performing a leak or integrity survey as the final stage of quality assurance should be considered. It is easier to repair leaks before the liner is placed in service. Electrical techniques are most frequently used where electric potential is applied between the top and bottom surfaces of the geomembrane over the geomembrane liner and the subgrade. A current flows through leaks in the geomembrane liner, and a survey probe is used to measure the electric potential gradients in the water over the geomembrane liner that identifies the unique signal at a hole location.

The technique has four basic requirements:

- a conductive medium over the geomembrane liner,
- a conductive medium immediately under the geomembrane liner,
- a conductive medium through the leak passages, and
- isolation of the conductive media above and below the liner except through the leaks to be found.

For optimum leak/integrity survey sensitivity, the number of underwater metal or concrete penetrations through the liner, metal batten strips and bolts, and other conductive penetrations through the liner should be minimized.

## Solvent Testing

If a geomembrane liner system requires a solvent prewash, and/or a solvent-based adhesive seaming procedure is used, then the water must be tested for traces of residual solvent before returning the reservoir to service. After the geomembrane liner installation is complete and sufficient time has elapsed for evaporation of any residual solvent from the seaming process, the reservoir should be partially filled and flushed until testing for residual solvent reveals that an acceptable level has been reached. Normally, two to three cycles are adequate.

This page intentionally blank.

# Chapter 5

# Operation, Inspection, and Maintenance Guidelines for Geomembrane Liners

## GENERAL

The purpose of these guidelines is to assist the water system operator in developing a sound operation and maintenance (O&M) plan for any potable water reservoir with a geomembrane liner system. Such a plan will, when developed, minimize the cost of operation and optimize the service the facility provides.

## OPERATION

### General Guidelines

**Reservoir startup.** After the installation is complete and any residual solvent used in the seaming process evaporates or has been flushed out, the filling of the reservoir may begin. A reservoir-specific fill plan should be developed with flow rates, durations, reservoir water levels, and total volume of water. The fill plan should be coordinated with the reservoir disinfection plan.

At low operating levels, the geomembrane liner is susceptible to damage by water entering the reservoir at high flows. The initial fill should commence at reduced flows until the potential for damage from water impinging onto the geomembrane liner, seams, and attachment points is greatly reduced. The geomembrane liner should be visually observed during the fill procedure until the entire reservoir floor is covered with several feet of water and all inlet facilities are fully submerged. The geomembrane liner should continue to be observed, but at a reduced degree, until the water reaches its maximum normal operating level and the reservoir is placed into normal operational mode. This precaution is necessary to ensure that air is not trapped between the geomembrane liner and the substrate, which would cause undesirable stresses to the geomembrane liner.

**Disinfection.** Disinfection of the reservoir should be performed by the owner. The reservoir should be disinfected after the completion of the initial geomembrane liner installation and after removing the reservoir from service for cleaning or repairs. The most common disinfection method for geomembrane liner installations is ANSI/AWWA C652, Disinfection of Water-Storage Facilities, Method 3. The suitability of the disinfection method and concentration of chemicals should be confirmed by the geomembrane material manufacturer to prevent potential damage to the geomembrane liner material and system components. The main disinfection objective should be to provide a free residual chlorine level of 2 mg/L after 24 h contact time and a bacteria-free condition in the entire reservoir volume before returning it to service. The required amount of disinfectant should be added to the reservoir at low water level or preferably introduced into the incoming water lines. The disinfection procedure should be submitted and approved by the applicable state regulatory agencies. Water quality tests for bacteria, organics, and inorganics should be performed before returning the reservoir to service.

**Dewatering.** Whenever a reservoir is being dewatered, the geomembrane liner should be visually observed. Groundwater or leakage from existing structures may complicate these operations. A reservoir-specific dewatering plan should be developed stating outflow rates, durations, reservoir water levels, and total volume of water. The dewatering plan should be based on site-specific groundwater and geotechnical conditions.

Ideally, the water surface in the reservoir should be greater than the groundwater level. This condition should be closely monitored during dewatering operations, and the dewatering rate may need to be adjusted to satisfy this recommendation. When empty, the geomembrane liner is subject to damage from groundwater pressures caused by saturated soil conditions behind or beneath the geomembrane liner. The potential for such conditions should be anticipated in the reservoir design, and adequate relief valves or an underdrain system should be provided to relieve excessive pressure. Where water buildup occurs, valves should be opened or the geomembrane liner slit to relieve the undesirable pressure. Relief valves should be closed and slits repaired before the reservoir is refilled. If relief valves are used, they should be designed to prevent contaminants from entering the reservoir while it is in operation.

**Cold-weather considerations.** Depending on the type of reservoir and its location, configuration, and use, a layer of surface ice can cause serious abrasion where it sustains contact with the geomembrane liner material at the side slopes as the water surface level falls. Such abrasion can be extremely detrimental to the geomembrane liner material and seams. To avoid excessive wear on the geomembrane liner, water-level fluctuations should be minimized as much as practical when surface ice conditions exist. Because of the distress that the geomembrane liner can be subjected to by reservoir icing, it is recommended that the reservoir be examined for damage after each winter season.

# INSPECTION

This section addresses inspection procedures recommended for reservoirs without floating covers. Alternative inspection procedures are required for inspection of installations with floating covers. As discussed previously, a reservoir geomembrane liner is subject to stress created by subgrade settlement, water flowing over the geomembrane liner surface, water-level changes, and ice movement—all of which can cause geomembrane material splits, seam failures, and anchorage failures. To prevent excessive damage from such failures when a reservoir is placed in operation, visual inspection of the geomembrane liner system should be made during the first fill. Thereafter, throughout the life of the reservoir, regular inspections should be performed where possible. If failures are observed or suspected, a more detailed inspection should be conducted. If necessary, the reservoir should be taken out of service to determine whether failures exist. If failures are discovered, the causes should be determined and corrected before returning the reservoir to service.

The reservoir may require periodic dewatering to allow cleaning and inspection of the entire geomembrane liner system. Dewatering should be done slowly such that any leakage behind the liner can dissipate at the same rate as the water over the liner. After the reservoir is dewatered and cleaned, the entire geomembrane should be visually inspected as follows:

- All material should be inspected for punctures, evidence of leakage, failure or excess settlement of substrate beneath the geomembrane liner, and material degradation. Where the integrity of the material is questioned, the geomembrane material manufacturer, engineer, or testing laboratory should be consulted.
- All factory and field seams should be inspected visually for separation or bond failure and appropriate repairs performed. Additional vacuum tests may be performed on suspect seams.
- All mechanical anchorages should be checked for pullout, distress, or leakage. When detected, the cause of the failure should be analyzed carefully and corrected before repairs are made.
- All joints bonding the liner to appurtenances (such as columns, pipe penetrations, gauge lines, and valves) should be carefully checked and reconnected where necessary.
- Bridging of geomembrane liner material should be investigated at the toe of slope, transitions, anchorage points, and all other potential bridging locations.

## Underwater Inspection

For reservoirs with a geomembrane liner that cannot be drained and that do not have a liner-leakage collection system and an underdrain system, an underwater inspection by divers or ROVs should be performed every other year. Reservoirs with a geomembrane liner, a liner-leakage collection system, and underdrain system do not require regular underwater inspections. For reservoirs with floating covers, underwater inspections of the geomembrane liner shall be performed at the same time and frequency as that required for the floating cover. Additional underwater inspections may be performed on an as-needed basis to investigate specific concerns regarding potential damage to the geomembrane liner. The reservoir should ideally be in the near-full condition for underwater inspections. The owner may consider additional underwater inspections after heavy winters with the potential for or suspicion of damage from freezing conditions. The procedures outlined in Chapter 3 should be used.

All underwater inspections shall be videotaped or photographed. The sequence of inspection shall be scripted by the owner and followed for every periodic underwater inspection to provide a constant and uniform comparison with previous inspections.

### Geomembrane Liner Leakage

The geomembrane liner underdrain system should be regularly checked in accordance with the operations and maintenance plan. The leakage rate should be recorded for historical tracking purposes. The operations and maintenance plan should identify acceptable leakage rates and identify the threshold leakage rate where investigative procedures should be performed.

## MAINTENANCE

### Cleaning

After the reservoir is emptied, a typical geomembrane liner system can be cleaned using a fire hose connected to the potable water supply. Extreme care should be taken to avoid spraying water at high pressure (above 40 psi [275 kPa]) or through high-velocity nozzles, since these practices could open seams or extend existing openings or splits. Wet geomembrane liner material can be extremely slick, so adequate safety precautions should be taken to protect employees from slips and falls during washing operations.

### Geomembrane Repairs

Geomembrane liner repairs should be performed only by properly trained personnel using materials and methods outlined in the operations and maintenance manual. Repair work should only be performed on dry geomembrane material. Because taste- and odor-causing chemicals may be used in making the repairs, extreme caution must be taken to ensure that any such chemicals are thoroughly dissipated once repairs are completed.

### Failures

Wherever a tear, rip, seam failure, or similar undesirable breach is found in a geomembrane liner, repairs should be made immediately after completing a thorough investigation of the cause of the failure. Such investigations are necessary to preclude a recurrence of the failure after repairs are made and to avoid similar failures in other areas of the reservoir. All repairs should be made in strict accordance with the recommendations of the operations and maintenance manual. Failures may be cut out for archiving and future examination.

### Bond and Anchorage Failures

When adhesives or mechanical anchorages fail to seal the geomembrane liner material in the manner originally designed, performance of such systems should be investigated. Investigation of the cause of failure is essential to determine if similar bond or anchorage failures can be expected in other areas of the reservoir. Investigation should also determine if a permanent or temporary repair is appropriate. Upon completion of the investigation and before returning the reservoir to service, all bond and anchorage failures should be repaired. In addition, repairs should strengthen all joints in which the investigation reveals a likelihood of further failures. All repairs should be made in strict accordance with the recommendations of the operations and maintenance plan.

## System Modification

Investigations of the causes of geomembrane liner failure may indicate the need to redesign or modify the system to avoid a recurrence. Such modifications should be made before the reservoir is returned to service. Temporary repairs may be made—provided the geomembrane liner system is monitored frequently while in service—pending permanent modifications and repairs.

This page intentionally blank.

# Bibliography

American National Standards Institute/American Water Works Association (ANSI/AWWA). C652. Disinfection of Water-Storage Facilities. Denver: AWWA.

ANSI/AWWA. D130. Geomembrane Materials for Potable Water Applications. Denver: AWWA.

ASTM. D4439. Standard Terminology for Geosynthetics. West Conshohocken, Pa.: ASTM.

AWWA Committee on Concrete Water Holding Structures. 1978. "A Summary Report on Concrete Water-Holding Structures." *Journal AWWA*. 70(8):458.

AWWA Committee on Control of Water Quality in Transmission and Distribution Systems. 1983. "Deterioration of Water Quality in Large Distribution Reservoirs (Open Reservoirs)." Committee Report. *Journal AWWA*. 75(6):313.

AWWA California–Nevada Section. 1999. "Reservoir Floating Cover Guidelines." Rancho Cucamonga, Calif.: AWWA California–Nevada Section.

Bailey, S.W. and E.C. Lippy. 1978. "Should All Finished Water Reservoirs Be Covered?" *Public Works*. 109(4):66.

Banta, J. 1968. "Our Reservoir Cover Floats on Water." *The American City*. 83(7):125.

Brown, J.H. 1971. "Flexible Membrane: An Economic Reservoir Liner and Cover." *Journal AWWA*. 71(6):328.

Bumstead, J.H. 1976. "Seamless Leakproof Lining for Linden Reservoir." *Journal AWWA*. 68(11):571.

"Butyl Membrane for Indian Reservoir." 1971. *Rubber Age*. 103(12):75.

Chack, R.T. 1970. "Largest Butyl Rubber Lined Reservoir." *Civil Engineering*. 40:44 (May).

Chin, A.G. 1971. "Covering Open Distribution Reservoirs." *Journal AWWA*. 63(12):763.

Crow, F.R. 1973. "Increasing Water Supplies by Suppression of Reservoir Evaporation." Oklahoma Water Resources Research Institute. Stillwater, Okla.

Culp, R., et al. 1984. "Heavy Snow Loads Plague Reservoir Roof." *Public Works*. 115(8):80.

Dallaire, E.E. 1975. "The Floating Cover: Best Way to Cover a Finished-Water Reservoir." *Civil Engineering*. 45(6):75.

"Economical Membrane Covers Potable Water Reservoir." 1985. *Journal AWWA*. 77(12):91.

"EPDM Elastomers." 1973. *Rubber Age*. 105(9)43.

"80 Acres of PVC Saves City's Water Supply." 1970. *Construction Methods*. 52(2):101.

Elwell, F.H. 1979. "Flexible Reservoir Covers: A Case Study." *Journal AWWA*. 71(4):210.

——. 1976. "Flexible Covers for Reservoirs." Proc. AWWA Ann. Conf., New Orleans, La.

Gerber, D.H. 1984. "Floating Reservoir Cover Designs." Proc. Intl. Conf. Geomembranes, Denver.

Gerber, D.H., and R.M. Woodley. 1983. "Floating Covers for Potable Water Reservoirs—State of the Art." Proc. Intl. Sym. Wtr. Qual., Paris, France.

Geren, J.L. 1952. Salem Improves Water Supply With 100-MG Asphalt Lined Reservoir. *West. Construc*. 27:9:57 (Sept.).

Giroud, J.P., K. Badu-Tweneboah, and R. Bonaparte. 1992. "Rate of Leakage Through a Composite Liner Due to Geomembrane Defects." *Geotextiles and Geomembranes*. 11(1):1.

Giroud, J.P., M.V. Khire, and K.L. Soderman. 1997. "Liquid Migration Through Defects in a Geomembrane Overlain and Underlain by Permeable Media." *Geosynthetics International*. 4(3–4):293.

Harem, F.E., K.D. Bielman, and J.E. Worth. 1976. "Reservoir Linings." *Journal AWWA*. 68(5):238.

Hovater, L.R. 1973. "No Water Lost From This Reservoir." *Public Works*. 104:72 (Mar.).

Hypalon™ Lining/Cover Restores City Reservoirs in Philadelphia. 1976. *Rubber World*. 174:53 (April).

Kays, W.B. 1977. *Construction of Linings for Reservoirs, Tanks, and Pollution Control Facilities*. New York: John Wiley & Sons.

——. 1977. "Cleaning Up Some Misconceptions About Basic Lining Technology." *Water & Sewage Works*. 124:10:52.

——. 1977. "Roof System Controls Pollution in Reservoirs." *Water & Sewage Works*. 124:11:75 (Nov.).

Kirby, P.C. 1979. "The Selection of Membrane Materials for Water Distribution and Allied Applications." Tech. Rept., Wtr. Res. Centre, United Kingdom.

Kittredge, D. 1970. "Floating Covers for Water Reservoirs." *Journal New Hampshire Water Works Association*. 31:5:13 (Oct.).

——. 1971. "Effects of Covering Water Reservoirs." *Water & Wastes Engineering*.

——. 1984. "Performance of Flexible Membrane Floating Covers." Proc. Intl. Conf. Geomembranes, Denver.

——. 1984. "Design and Operation of Flexible Membrane Floating Covers for Water Distribution Reservoirs." *Journal NEWWA*. 98:3:2590.

Kittredge, D., and R. Taylor. 1993. "Flexible Floating versus Concrete Covers for Water Storage." *Public Works*. (Nov.)

Kittredge, D., and R. Beaurivage. 1994. "Experience With Floating Covers at the Manchester Water Works." Proc. AWWA Ann. Conf., New York, N.Y.

Krasner, S., and E. Means III. 1986. "Returning Recently Covered Reservoirs to Service: Health and Aesthetic Considerations." *Journal AWWA*. 78(3):94.

"Largest Butyl Lined Reservoir." 1974. *Public Works*. 105:92 (Mar.).

"Leaky Reservoir Stopped With Liquid PU Rubber." 1974. *Rubber World*. 170:52 (June).

Lee, J. 1974. "Selecting Membrane Pond Liners." *Poll. Engrg*. 6:1:35 (June).

"Liner and Floating Cover Protect California Reservoir." 1985. *Public Works*.

Maddocks, D. 1975. *Methods of Creating Low-Cost Waterproof Membranes for Use in the Construction of Rain-Water Catchment and Storage Systems*. London: Intermediate Technology Publications.

"Million Gallon Fabric Water Tank Performing Well." 1973. *Public Works*. 104:120 (June).

More, L.L. 1979. "View Point: Open Finished Water Reservoirs." *Journal AWWA*. 71(8):11.

NSF/ANSI/CAN 61. Drinking Water System Components—Health Effects. Washington, DC: ANSI.

"News of the Field—Floating Lids and Liners Protect Philadelphia Reservoir Water." 1977. *Journal AWWA*. 69(12):32.

"Spokane Solves Water Problem with Reservoir Liner and Cover." 1982. *Const. Specifier* (June).

Taylor, R.T., and L.R. Schader. 1993. "Geomembrane Floating Cover Technology for the Nineties." Proc. Geosynthetics Conf. Vancouver, B.C.

Uniroyal Water Tank. 1975. "New Flexible Tank Holds Potable or Process Water." *Canadian Min. & Met. Bull.*, 68:107D.

Wade, L.R. 1975. "PVC Makes Big Reservoir Better." *Water & Wastes Engineering*, 12:57 (Nov.).

"Water Tank." 1972. *Rubber World*. 166:21 (May).

"Watertight Butyl Rubber Sheet Lines Basins in Water Handling Project." 1974. *Rubber Age*. 106:48 (Aug.).

Wood, W.G., and H.E. Howell. 1970. "Giving New Life to a 50-Year-Old Reservoir." *Public Works*. 101:106 (April).

# Index

Note: *f.* indicates figure

## L



## M



## O



## P



## Q



## R



## S



## T

This page intentionally blank.

# AWWA Manuals

M1, *Principles of Water Rates, Fees, and Charges,* #30001
M2, *Instrumentation and Control,* #30002
M3, *Safety Management for Water Utilities,* #30003
M4, *Water Fluoridation Principles and Practices,* #30004
M5, *Water Utility Management,* #30005
M6, *Water Meters—Selection, Installation, Testing, and Maintenance,* #30006
M7, *Problem Organisms in Water: Identification and Treatment,* #30007
M9, *Concrete Pressure Pipe,* #30009
M11, *Steel Pipe—A Guide for Design and Installation,* #30011
M12, *Simplified Procedures for Water Examination,* #30012
M14, *Backflow Prevention and Cross-Connection Control: Recommended Practices,* #30014
M17, *Fire Hydrants: Installation, Field Testing, and Maintenance,* #30017
M19, *Emergency Planning for Water and Wastewater Utilities,* #30019
M20, *Water Chlorination/Chloramination Practices and Principles,* #30020
M21, *Groundwater,* #30021
M22, *Sizing Water Service Lines and Meters,* #30022
M23, *PVC Pipe—Design and Installation,* #30023
M24, *Planning for the Distribution of Reclaimed Water,* #30024
M25, *Flexible-Membrane Covers and Linings for Potable-Water Reservoirs,* #30025
M27, *External Corrosion Control for Infrastructure Sustainability,* #30027
M28, *Rehabilitation of Water Mains,* #30028
M29, *Water Utility Capital Financing,* #30029
M30, *Precoat Filtration,* #30030
M31, *Distribution System Requirements for Fire Protection,* #30031
M32, *Computer Modeling of Water Distribution Systems,* #30032
M33, *Flowmeters in Water Supply,* #30033
M36, *Water Audits and Loss Control Programs,* #30036
M37, *Operational Control of Coagulation and Filtration Processes,* #30037
M38, *Electrodialysis and Electrodialysis Reversal,* #30038
M41, *Ductile-Iron Pipe and Fittings,* #30041
M42, *Steel Water-Storage Tanks,* #30042
M44, *Distribution Valves: Selection, Installation, Field Testing, and Maintenance,* #30044
M45, *Fiberglass Pipe Design,* #30045
M46, *Reverse Osmosis and Nanofiltration,* #30046
M47, *Capital Project Delivery,* #30047
M48, *Waterborne Pathogens,* #30048
M49, *Quarter-Turn Valves: Head Loss, Torque, and Cavitation Analysis,* #30049
M50, *Water Resources Planning,* #30050
M51, *Air Valves: Air-Release, Air/Vacuum and Combination,* #30051
M52, *Water Conservation Programs—A Planning Manual,* #30052
M53, *Microfiltration and Ultrafiltration Membranes for Drinking Water,* #30053
M54, *Developing Rates for Small Systems,* #30054
M55, *PE Pipe—Design and Installation,* #30055
M56, *Nitrification Prevention and Control in Drinking Water,* #30056
M57, *Algae: Source to Treatment,* #30057
M58, *Internal Corrosion Control in Water Distribution Systems,* #30058
M60, *Drought Preparedness and Response,* #30060
M61, *Desalination of Seawater,* #30061
M62, *Membrane Applications for Water Reuse,* #30062
M63, *Aquifer Storage and Recovery,* #30063

M65, *On-Site Generation of Hypochlorite,* #30065

M66, *Cylinder and Vane Actuators and Controls—Design and Installation,* #30066

M68, *Water Quality in Distribution Systems,* #30068

M69, *Inland Desalination and Concentrate Management,* #30069

M77, *Condition Assessment of Water Mains,* #30077